Asymmetry in Biological Homochirality

Asymmetry in Biological Homochirality

Editor

David Hochberg

MDPI • Basel • Beijing • Wuhan • Barcelona • Belgrade • Manchester • Tokyo • Cluj • Tianjin

Editor
David Hochberg
Centro de Astrobiología
(CSIC-INTA)
Spain

Editorial Office
MDPI
St. Alban-Anlage 66
4052 Basel, Switzerland

This is a reprint of articles from the Special Issue published online in the open access journal *Symmetry* (ISSN 2073-8994) (available at: https://www.mdpi.com/journal/symmetry/special_issues/Asymmetry_Biological_Homochirality).

For citation purposes, cite each article independently as indicated on the article page online and as indicated below:

LastName, A.A.; LastName, B.B.; LastName, C.C. Article Title. *Journal Name* **Year**, *Volume Number*, Page Range.

ISBN 978-3-0365-0442-1 (Hbk)
ISBN 978-3-0365-0443-8 (PDF)

Contents

About the Editor vii

Preface to "Asymmetry in Biological Homochirality" ix

Victor V. Dyakin, Thomas M. Wisniewski and Abel Lajtha
Chiral Interface of Amyloid Beta (Aβ): Relevance to Protein Aging, Aggregation and Neurodegeneration
Reprinted from: *Symmetry* **2020**, *12*, 585, doi:10.3390/sym12040585 1

Dilip Kondepudi and Zachary Mundy
Spontaneous Chiral Symmetry Breaking and Entropy Production in a Closed System
Reprinted from: *Symmetry* **2020**, *12*, 769, doi:10.3390/sym12050769 11

Carsten Tschierske and Christian Dressel
Mirror Symmetry Breaking in Liquids and Their Impact on the Development of Homochirality in Abiogenesis: Emerging Proto-RNA as Source of Biochirality?
Reprinted from: *Symmetry* **2020**, *12*, 1098, doi:10.3390/sym12071098 25

William Bock and and Enrique Peacock-Lopez
Chiral Oscillations and Spontaneous Mirror Symmetry Breaking in a Simple Polymerization Model
Reprinted from: *Symmetry* **2020**, *12*, 1388, doi:10.3390/sym12091388 55

Josep M. Ribó
Chirality: The Backbone of Chemistry as a Natural Science
Reprinted from: *Symmetry* **2020**, *12*, 1982, doi:10.3390/sym12121982 69

About the Editor

David Hochberg earned an undergraduate degree in physics from the University of California, Berkeley and continued his studies in theoretical physics, earning both a Masters in Science and a Ph.D. in physics from the University of Chicago. After holding postdoctoral research positions in the UK (Rutherford Appleton Laboratory), the USA (Bartol Research Institute), and the University of Vanderbilt), and in Spain (University of Valencia, Autonomous University of Madrid), he was appointed a permanent Scientific Researcher in the Spanish Science Research Council (CSIC) and is a founding member of the Centro de Astrobiología (CAB). He has more than thirty years of experience in applying methods of theoretical physics to a wide variety of problems that span general relativity, electrostatic fields of DNA in solution, and stochastic processes in chemistry and physics, among many others. His current research interests involve symmetry breaking processes in the physics and chemistry of complex systems, with an emphasis on mirror symmetry breaking in chemistry and in complex chemical systems. He is an expert in applying methods of quantum and stochastic field theory, the dynamic renormalization group, and statistical physics to nonlinear dynamical systems far from equilibrium. Dr. Hochberg served as head of the Chirality Working Group during the COST Action CM0703 "Systems Chemistry".

Preface to "Asymmetry in Biological Homochirality"

The chemistry of life on Earth is based on a basic asymmetry of certain molecules whose three-dimensional geometrical structure or conformation is not identical to that of their mirror image or spatial reflection through a mirror. We say that parity P, or space inversion, a fundamentally discrete spatial symmetry transformation of fundamental physics, is broken at the molecular level. Such molecules are said to possess chirality or handedness. The mirror image structures of a chiral molecule are called enantiomers. Homochirality is ubiquitous in biological chemistry from its very start. Amino acids, the building blocks of proteins, and the sugar backbones present in DNA and RNA, are chiral molecules. Thus, the question arises: What are the reasons for molecular systems to break their mirror symmetry? Furthermore, under what conditions? What physico-chemical mechanisms are required so that a tiny excess of one enantiomer leads to chiral amplification at the macroscopic level? What is the origin of homochirality on Earth? Progress regarding these problems can only make great strides from the participation of researchers working on chiral symmetry, and coming from diverse backgrounds: Theory, experiment, non-equilibrium thermodynamics, crystallography, catalysis, nucleation, chemical engineering, liquid crystals, surface science, spectroscopy, organic synthesis, and quantum chemistry.

This book presents a selection of current cutting-edge chirality research by leading experts in the field. Dyakin, Wisniewski, and Laytha review protein chirality, which embraces biophysics and biochemistry. In this article, the authors focus attention on the amyloid-beta (Aβ) peptide, which is known for its essential cellular functions and associations with neuropathology. Kondepudi and Mundy present a study of a theoretical model of a photochemically driven, closed chemical system in which spontaneous chiral symmetry breaking occurs, and consider the entropy production before and after the chiral symmetry breaking transition. Tschierske and Dressel thoroughly review mirror symmetry breaking in liquids by tracking the recent progress in mirror symmetry breaking and chirality amplification in isotropic liquids and liquid crystalline cubic phases of achiral molecules and discussing its implications for the hypothesis of emergence of biological chirality. Bock and Peacock-López revisit the activation–polymerization–epimerization–depolymerization (APED) model, originally proposed to describe chiral symmetry breaking in a simple dimerization system. They extend APED and consider the role of higher oligomers, from trimers to pentamers, for chiral and chemical oscillations that exist for certain system parameters and reveal the preferential formation of heterochiral polymers that results. Finally, Ribó considers the role of chiral dissymmetry in the chemical evolution towards life and how the increase of chemical complexity, from atoms and molecules up to complex open systems, accompanies the emergence of biological homochirality.

David Hochberg
Editor

Article

Chiral Interface of Amyloid Beta (A*β*): Relevance to Protein Aging, Aggregation and Neurodegeneration

Victor V. Dyakin [1,*], Thomas M. Wisniewski [2] and Abel Lajtha [1]

[1] Departmemts: Virtual Reality Perception Lab. (VV. Dyakin) and Center for Neurochemistry (A. Lajtha), The Nathan S. Kline Institute for Psychiatric Research (NKI), Orangeburg, NY 10962, USA; Abel.Lajtha@NKI.rfmh.org

[2] Departments of Neurology, Pathology and Psychiatry, Center for Cognitive Neurology, New York University School of Medicine, New York, NY 10016, USA; thomas.wisniewski@nyulangone.org

* Correspondence: Victor.dyakin@nki.rfmh.org

Received: 17 March 2020; Accepted: 2 April 2020; Published: 7 April 2020

Abstract: Biochirality is the subject of distinct branches of science, including biophysics, biochemistry, the stereochemistry of protein folding, neuroscience, brain functional laterality and bioinformatics. At the protein level, biochirality is closely associated with various post-translational modifications (PTMs) accompanied by the non-equilibrium phase transitions (PhTs ^{NE}). PTMs ^{NE} support the dynamic balance of the prevalent chirality of enzymes and their substrates. The stereoselective nature of most biochemical reactions is evident in the enzymatic (Enz) and spontaneous (Sp) PTMs (PTMs ^{Enz} and PTMs ^{Sp}) of proteins. Protein chirality, which embraces biophysics and biochemistry, is a subject of this review. In this broad field, we focus attention to the amyloid-beta (Aβ) peptide, known for its essential cellular functions and associations with neuropathology. The widely discussed amyloid cascade hypothesis (ACH) of Alzheimer's disease (AD) states that disease pathogenesis is initiated by the oligomerization and subsequent aggregation of the Aβ peptide into plaques. The racemization-induced aggregation of protein and RNA have been extensively studied in the search for the contribution of spontaneous stochastic stereo-specific mechanisms that are common for both kinds of biomolecules. The failure of numerous Aβ drug-targeting therapies requires the reconsolidation of the ACH with the concept of PTMs ^{Sp}. The progress in methods of chiral discrimination can help overcome previous limitations in the understanding of AD pathogenesis. The primary target of attention becomes the network of stereospecific PTMs that affect the aggregation of many pathogenic agents, including Aβ. Extensive recent experimental results describe the truncated, isomerized and racemized forms of Aβ and the interplay between enzymatic and PTMs ^{Sp}. Currently, accumulated data suggest that non-enzymatic PTMs ^{Sp} occur in parallel to an existing metabolic network of enzymatic pathways, meaning that the presence and activity of enzymes does not prevent non-enzymatic reactions from occurring. PTMs ^{Sp} impact the functions of many proteins and peptides, including Aβ. This is in logical agreement with the silently accepted racemization hypothesis of protein aggregation (RHPA). Therefore, the ACH of AD should be complemented by the concept of PTMs ^{Sp} and RHPA.

Keywords: biochirality; post-translational modifications; protein folding; protein aggregation; spontaneous chemical reactions; neurodegeneration; non-equilibrium phase transitions

---

1. Introduction

The amyloid cascade hypothesis (ACH) has played a crucial role in the understanding of the Alzheimer's disease (AD) etiology and pathogenesis. The deposition of β-amyloid (Aβ) and neurofibrillary tangles (NFTs) traditionally served as the essential neuropathological features of AD. However, for many years, the attention to the stereochemistry of underlying spontaneous events was under-appreciated. Stereochemical errors in biomolecular structures, including proteins and peptides, have a dramatic impact on cell physiology [1]. The discovery of free D-aspartic acid (D-Asp) in rodents and man open a new window for understanding the mechanisms of protein synthesis and degradation [2]. Proteins, including glycoproteins, are the subjects of the reversible enzymatic (Enz) [3] and irreversible spontaneous (Sp) [4] post-translational modifications (PTM ^{Enz} and PTMs ^{Sp}). (see Figure 1). All physiological and pathological forms of proteins are the consequence of PTMs. We are focusing on the aberrant forms of PTM, such as racemization ^{Sp} and isomerization ^{Sp}. The relevance of the spontaneous modifications of amino acids (AAs) within peptides and long-lived proteins to protein aging, accumulation and pathologies is being increasingly recognized in the recent studies. Accordingly, the primarily biomarkers of aging and neurodegeneration are becoming the protein-cell-specific PTMs ^{Sp} of amino acids (AAs) [5–50].

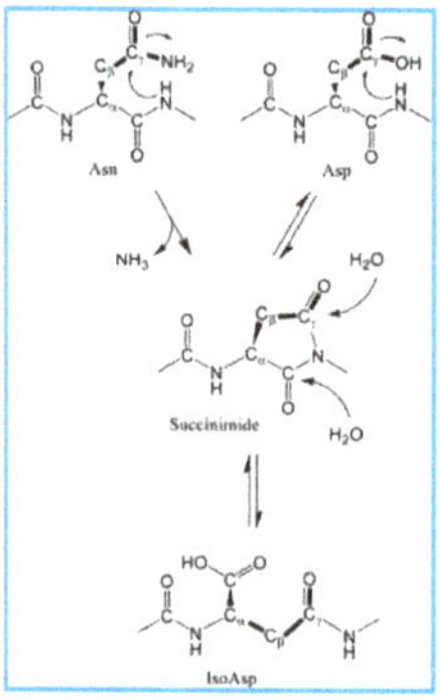

Figure 1. Spontaneous deamidation and isomerization of asparagine (Asn). Side-chain bonds of asparagine and aspartate are drawn as bold lines. Adopted from [4].

2. Racemization of the Aβ

With the recognition of the fact that many proteins (Aβ, TAU, prion protein Prion (PrP), Huntingtin and α-synuclein) are the substrate of the aggregation-prone PTMs ^{Sp} [51], we are focusing, primarily, on the racemization of the Aβ. The amyloid precursor protein (APP) is one of the most studied proteins concerning pathological misfolding. The products of APP processing by α-, β- and γ-secretases range from 16 to 49 AAs.

Most studied amyloid beta (Aβ) peptides includes Aβ (1–16), Aβ (1–19), Aβ (20–34), Aβ (20–33), Aβ (20–40), Aβ (23–34), Aβ (34–40), Aβ (35–40), Aβ (1–35), Aβ (1–40) and Aβ (1–42) [5,52] are characterized by differential chain of PTMs and susceptibility to PTMs ^{Sp}. PTMs of Aβ (1–42), the primary constituent of Aβ plaques in the AD brain, are extensively studied. Misfolding and aggregation of Aβ peptides is the convincing example of a link between the perturbations of the molecular chirality, deteriorated enzyme-substrate recognition, abnormal cell signaling (including neurotransmission) and cognitive dysfunction [3]. The spontaneous aggregation of Aβ peptides into amyloid plaques and in the walls of the cerebral vasculature is the unresolved issue of Alzheimer's disease (AD)-amyloid conundrum [6]. It is a common assumption that PTMs ^{Sp} can significantly alter the structure of the original polypeptide chain. The AAs that most frequently undergo racemization ^{Sp} and isomerization ^{Sp} in human proteins are aspartate (Asp), asparagine (Asn), glutamate (Glu), glutamine (Glu), serine (Ser), alanine (Ala) and proline [7]. For Aβ peptides, racemization-prone are found two non-essential AAs: serine (Ser) and aspartate (Asp) (Aβ-42 contains two Ser and three Asp residues (see Table 1)). For Asp, the mechanism acceleration of racemization ^{Sp} (about 10^{5}) is associated with the specific succinimide intermediates [8–10]. Both D-Ser and D-Asp play a crucial role in N-methyl-d-aspartate (NMDA) receptor-mediated neurotransmission. D-Ser26-Aβ1–40 possesses a strong tendency to form fibrils [11]. AD patients have increased brain D-Ser levels [12]. This fact agrees with the activated spontaneous racemization (Rs^{Sp}) of Ser residue in Aβ, with an elevated level of D-Ser in amyloid plaques, impairment of the NMDA neurotransmission, memory loss and cognitive dysfunction. Racemization and isomerization of Asp are the most common types of non-enzymatic covalent modification that leads to an accumulation of aging proteins in numerous human tissues [13]. Asp-1, Asp-7 and Asp-23 of Aβ are crucial in the control of Aβ aging and aggregation [5].

Table 1. The frequency (f) of the AAs appearance in Aβ (1–42).

The Frequency (f) of the AAs Appearance in A-beta (1–42)					
f			Amino Acids		
6	Gly	Val			
4	Ala				
3	**Asp**	Glu	Phe	His	Ile
2	Lys	Leu	**Ser**		
1	Met	Asn	Arg	Gln	Tyr

Residues Asp-1 and Asp-7 of Aβ in amyloid plaques are a mixture of L-, D-, L-iso- and D-iso-aspartate [14]. D-Asp-7 enhances the aggregation process by shifting the equilibrium of Aβ from the soluble to the insoluble form [15]. Therefore, the set of PTMs ^{Sp}, including racemization ^{Sp} and isomerization ^{Sp}, is an efficient modifier of Aβ metabolism. In 2011, Kumar promoted the hypothesis that enzymatic phosphorylation of Aβ triggers the formation of toxic aggregates [16], which has been confirmed by later studies [3]. In 1994, Szendrei discovered that spontaneous isomerization of Asp affects the conformations of synthetic peptides [17]. However, the role of the PTMs ^{Sp} in Aβ aggregation and neurotoxicity remains in the shadow. Consequently, many structural details of misfolded Aβ have remained elusive for a long time [7,18]. This short review provides a summary of information regarding events of PTMs ^{Sp} in Aβ. The heterogeneity of Aβ proteolytic forms in AD brain is represented by at least 26 unique peptides, characterized by various N- and C-terminal truncations. The N- and C-terminal truncated fragments (in contrast to canonical Aβ) are allowing to distinguish between the soluble and insoluble aggregates. The N-terminal truncations are predominating in the insoluble material and C- terminal truncations segregating in the soluble aggregates [19,20]. Aβ peptides exhibit a high sensitivity of the secondary structure and fibril morphologies to the chirality of ligands [21] and enzymes of PTMs. Only for a small part of Aβ isoforms exist information regarding the pathways of PTMs ^{Sp}.

Currently, available data for Aβ42 peptides are summarized in Tables 1–3. The data in Tables 2 and 3 demonstrate two essential facts: first, the coincidence of phosphorylation and spontaneous racemization/isomerization (enzymatic phosphorylation can be accompanied by the enzyme-driven or spontaneous racemization) events at the Ser-8, Ser-26 (Ser-26 residue is located within the turn region of Aβ), and Asp23 residues, second, currently available information covering the PTMs ^{Sp} of Aβ is limited only to 4 from the 16 types of AAs, which means that much remains to be explored. Most recent attempts to overcome previous limitations of the ACH are concentrated on many additional essential aspects of metabolism, contributing to progress in understanding [22–42]. However, most them, do not pay enough attention to the stereochemistry of the PTM in general and the impact of spontaneous racemization (^{Sp}) on Aβ assembly, aggregation and functions. At the same time, the progress in methods of chiral discrimination has produced new, stereochemistry-oriented, experimental results regarding aberrant PTMs ^{Sp}. Growing evidence suggests that proteins undergo several unusual, previously unknown PTMs associates with the interplay between physiological protein modification, spontaneous aging-associated molecular processes [7,10,11,13], stress conditions [23,24], accidental co-localization of the enzyme and substrate or PTMs ^{Sp}. For Aβ, Asp and Ser are known as the most racemization prone residues. For the illustration purpose, we provide several of many existing molecular pathways where the racemization ^{Sp} of Ser can be critical.

Table 2. Coincidence of enzymatic and spontaneous PTMs at Ser and Asp residues of Aβ in the neurodegenerative amyloid aggregates.

Peptide	Disease	Residue	PTMs		
			Rcm.	**Ism.**	**Ph.**
			Spontaneous		**Enzymic**
A-β (40–42)	AD	**Ser-8**	[24, 35]		[3, 40, 41, 42]
		Ser-26	[13, 35, 36]		[43. 44]
		Asp-23	[35, 37, 38]		[40]
A-β (20–34)		**Asp-23**		[39]	

Post-translational modification (PTMs): Spontaneous (Sp) and Enzymic (Enz), Racemization (Rcm). Isomerization (Ism). Phosphorylation (Ph).

Table 3. Coincidence of phosphorylation and racemization at Ser and Asp residues of Aβ (1–42).

	A-Beta (1–42)																																										
N-Terminal	1	2	3	4	5	6	7	8	9	10	11	12	13	14	15	16	17	18	19	20	21	22	23	24	25	26	27	28	29	30	31	32	33	34	35	36	37	38	39	40	41	42	C-Terminal
	Asp	Ala	Glu	Phe	Arg	His	Asp	**Ser**	Gly	Tyr	Glu	Val	His	His	Gln	Lys	Leu	Val	Phe	Phe	Ala	Glu	**Asp**	Val	Gly	**Ser**	Asn	Lys	Gly	Ala	Ile	Ile	Gly	Leu	Met	Val	Gly	Gly	Val	Val	Ile	Ala	
	D	A	E	F	R	H	D	**S**	G	Y	E	V	H	H	Q	K	L	V	F	F	A	E	**D**	V	G	**S**	N	K	G	A	I	I	G	L	M	V	G	G	V	V	I	A	
								*																		*																	
																							**																				
								***															***			***																	

PTM: Enzymic (***) and Spontaneous (Racemization *, Izomerization **).

First, the pathological role of mitochondrial enzymes Ser proteases (SerPs) is attributed to neurodegenerative disorders such as AD and Parkinson's and disease [25–27]. The HtrA (high-temperature requirement) family represents a class of oligomeric SerPs [25]. Its members are classified by presence (in its AAs sequence) a catalytic triad contains His, Asp and Ser residues known as racemization prone.

Second, chaperone signaling complexes in AD involve a wide range of heat shock proteins (Hsp), including Hsp27, that are ingaged in protection against Aβ aggregation and toxicity [28]. Human Hsp27 is phosphorylated at three Ser residues (Ser15, Ser78 and Ser82), were Ser-78 and Ser-82 are the major phosphorylation sites [29,30]. It is evident that due to the stereo-specificity activity of both protein types (SerPs and Hsps) racemization ^{Sp} of Ser residues in each of them will contribute to the aberrant processing of substrates, including Aβ, inducing the cascade of aggregation, accompanied by neurodegeneration. The aggregation of protein and peptide indicates the decrease in the turnover rate. Accordingly, the previously short half-life-proteins are changing in the direction toward the long-lived one. Lowering turnover rate (i.e., protein aging) makes proteins the subject of the time-dependent PTMs ^{Sp}, including oxidation, nitration, glycation, isomerization and racemization [7]. The set of PTMs ^{Sp} and its effect on protein polymerization both are substrate specific. Tyrosine (Tyr) nitration, for example, significantly decreased the aggregation of Aβ1–40. [43].

3. Conclusions

In the manuscript, we assess the previous and current experimental results acquired in the specific areas of chiral proteomics—Aβ folding—from a broad perspective. For this purpose, we addressed the basic, fundamental and widely recognized facts and theories underlying the stereochemistry of Aβ. Due to progress in multidisciplinary fields, the view of the origin of biologic non-equilibrium chirality evolves from the physico-chemical nature of enantioselective autocatalytic reaction networks to a process that play an essential role in the pathogenesis of AD [44]. The phenomena of biochirality embrace two undivided branches of science biophysics and biochemistry. In 1990th, the nature of living organisms was associated with the absolute homochirality [50]. With the discovery of D-AAs in living organisms and the process of enzymic racemization, the concept of homochirality was replaced by the notion of prevalent chirality.

In the language of entropy, the transfer of protein/solvent system from the state of low-entropy (racemic mixture) to the high-entropy state (homochirality) is the order-disorder type transitions.

In terms of thermodynamics, this is the transition from the non-equilibrium to the equilibrium state. Accordingly, the enzymic PTMs, from a biophysical perspective, is the set of physiological non-equilibrium phase transitions. Enzymic racemization is an essential and necessary source of D-AAs in organisms. In contrast, the spontaneous racemization, as an aberrant PTMs, is the window for the irreversible transfer from non-equilibrium to equilibrium conditions.

In the words of proteomics, irreversible racemization is the conformation of protein from functional (physiological) to the dis-functional (inert or toxic) state of protein solvent, aggregates and depositions. The universal significance of the symmetry constraints is evident from the viral to the human proteome. The biologic significance of racemization-induced protein aggregation for the neuropathogenesis of AD was experimentally demonstrated as early as 1994 1994 [45]. Currently accumulated data about PTMs of many proteins and peptides, including Aβ are coherent with the silently accepted racemization hypothesis of protein aggregation (RHPA). Therefore, after "three decades of struggles, ACH [46] of the neurodegeneration should be complemented by the concept of PTMs ^{Sp} and non-equilibrium phase transitions (PhTs ^{NE}) [47–50].

Author Contributions: V.V.D. contribute to review of biophysical aspect of molecular chirality, PTMs, protein aggregation, and neurodegeneration. A.L. contribute to review of biochemical aspect of molecular chirality, PTMs, protein aggregation, and neurodegeneration. T.M.W. contributed to review of biochemical aspects of PTMs, protein aggregation and neurodegeneration. All authors have read and agreed to the published version of the manuscript.

Funding: This manuscript is supported by NIH grants AG060882, AF058267 and AG066512 to TMW.

Acknowledgments: We gratefully thanks: Justin Lucas for criticism and correction. Csaba Vadasz for criticism and correction. Alexander G. Dadali for discussion of the thermodynamics of protein conformation, and Frank Fagnano for contribution to clarity of the argumentation.

Conflicts of Interest: The authors declare no conflicts of interest.

Abbreviations

ACH	Amyloid cascade hypothesis
PTMs	post-translational modifications
PTMs Enz	enzymic PTMs
PTMs Sp	spontaneous PTMs
PhTs NE	non-equilibrium phase transitions
Aβ	amyloid beta
RHPA	racemization hypothesis of protein aggregation
Ala	Alanine
Asn	Asparagine
Ser	Serine
Asp	aspartic acid
Glu	Glutamate
Ile	Isoleucine
Tyr	Tyrosine
Pro	Proline

References

1. Schreiner, E.; Trabuco, L.G.; Freddolino, P.L.; Schulten, K. Stereochemical errors and their implications for molecular dynamics simulations. *BMC Bioinform.* **2011**, *12*, 190. [CrossRef] [PubMed]
2. Dunlop, D.S.; Neidle, A.; McHale, D.; Dunlop, D.M.; Lajtha, A. The presence of free D-aspartic acid in rodents and man. *Biochem. Biophys. Res. Commun.* **1986**, *141*, 27–32. [CrossRef]
3. Jamasbi, E.; Separovic, F.; Hossain, M.A.; Ciccotosto, G.D. Phosphorylation of a full-length amyloid-β peptide modulates its amyloid aggregation, cell binding and neurotoxic properties. *Mol. BioSyst.* **2017**, *13*, 1545–1551. [CrossRef] [PubMed]
4. Yokoyama, H.; Mizutani, R.; Noguchi, S.; Hayashida, N. Structural and biochemical basis of the formation of isoaspartate in the complementarity-determining region of antibody 64M-5 Fab. *Sci. Rep.* **2019**, *9*, 18494. [CrossRef]
5. Moro, M.L.; Collins, M.J.; Cappellini, E. Alzheimer's disease and amyloid β-peptide deposition in the brain: A matter of 'aging'? *Biochem. Soc. Trans.* **2010**, *38*, 539–544. [CrossRef]
6. Roher, A.E.; Kokjohn, T.A.; Clarke, S.G.; Sierks, M.R.; Maarouf, C.L.; Serrano, G.E.; Sabbagh, M.S.; Beach, T.G. APP/Aβ structural diversity and Alzheimer's disease pathogenesis. *Neurochem. Int.* **2018**, *110*, 1–13. [CrossRef]
7. McCudden, C.R.; Kraus, V.B. Biochemistry of amino acid racemization and clinical application to musculoskeletal disease. *Clin. Biochem.* **2006**, *39*, 1112–1130. [CrossRef]
8. Radkiewicz, J.L.; Zipse, H.; Clarke, S.; Houk, K.N. Accelerated racemization of aspartic acid and asparagine residues via succinimide intermediates: An ab initio theoretical exploration of mechanism. *J. Am. Chem. Soc.* **1996**, *118*, 9148–9155. [CrossRef]
9. Helfman, P.M.; Bada, J.L.; Shou, M.Y. Considerations on the role of aspartic acid racemization in the aging process. *Gerontology* **1977**, *23*, 419–425. [CrossRef]
10. Takahashi, O.; Kirikoshi, R.; Manabe, N. Academic editor Mihai V. Putz. Racemization of the succinimide intermediate formed in proteins and peptides: A computational study of the mechanism catalyzed by dihydrogen phosphate ion. *Int. J. Mol. Sci.* **2016**, *10*, 1698. [CrossRef]

11. Kubo, T.; Kumagae, Y.; Miller, C.A.; Kaneko, I. Beta-amyloid racemized at the Ser26 residue in the brains of patients with Alzheimer's disease: Implications in the pathogenesis of Alzheimer's disease. *J. Neuropathol. Exp. Neurol.* **2003**, *62*, 248–259. [CrossRef] [PubMed]
12. Madeira, C.; Lourenco, M.V.; Vargas-Lopes, C.; Suemoto, C.K.; Brandão, C.O.; Reis, T.; Leite, R.E.P.; Laks, J.; Jacob-Filho, W.; Pasqualucci, C.A. D-serine levels in Alzheimer's disease: Implications for novel biomarker development. *Transl. Psychiatry* **2015**, *5*, e561. [CrossRef] [PubMed]
13. Ritz-Timme, S.; Collins, M.J. Racemization of aspartic acid in human proteins. *Ageing Res. Rev.* **2002**, *1*, 43–59. [CrossRef]
14. Roher, A.E.; Yablenson, J.D.; Clarke, S.; Wolkow, C.; Wang, R.; Cotter, R.J.; Reardon, I.M.; Zürcher-Neely, H.A.; Heinrikson, R.L.; Ball, M.J.; et al. Structural alterations in the peptide backbone of β-amyloid core protein may account for its deposition and stability in Alzheimer's disease. *J. Biol. Chem.* **1993**, *268*, 3072–3083.
15. Kuo, Y.M.; Webster, S.; Emmerling, M.R.; de Lima, N.; Roher, A.E. Irreversible dimerization/tetramerization and post-translational modifications inhibit proteolytic degradation of Aβ peptides of Alzheimer's disease. *Biochim. Biophys. Acta.* **1998**, *1406*, 291–298. [CrossRef]
16. Kumar, S.; Walter, J. Phosphorylation of amyloid beta (Aβ) peptides—A trigger for formation of toxic aggregates in Alzheimer's disease. *Aging* **2011**, *3*, 803–812. [CrossRef]
17. Szendrei, G.I.; Fabian, H.; Mantsch, H.H.; Lovas, S.; Nyéki, O.; Schön, I.; Otvos, L. Aspartate-bond isomerization affects the major conformations of synthetic peptides. *Eur. J. Biochem. FEBS* **1994**, *226*, 917–924. [CrossRef]
18. Xiao, Y.; Ma, B.; McElheny, D.; Parthasarathy, S.; Long, F.; Hoshi, M.; Nussinov, R.; Ishii, Y. Aβ (1–42) fibril structure illuminates self-recognition and replication of amyloid in Alzheimer's disease. *Nat. Struct. Mol. Biol.* **2015**, *22*, 499–505. [CrossRef]
19. Wildburger, N.C.; Esparza, T.J.; LeDuc, R.D.; Fellers, R.T.; Thomas, P.M.; Cairns, N.J.; Kelleher, N.L.; Bateman, R.J.; David, L.; Brody, D.L. Diversity of amyloid-beta proteoforms in the Alzheimer's disease brain. *Sci. Rep.* **2017**, *7*, 9520. [CrossRef]
20. Jiang, N.; Leithold, L.H.E.; Post, J.; Ziehm, T.; Mauler, J.; Gremer, L.; Cremer, M.; Schartmann, E.; Shah, N.J.; Kutzsche, J.; et al. Preclinical pharmacokinetic studies of the tritium labelled D-enantiomeric peptide D3 developed for the treatment of Alzheimer's disease. *PLoS ONE* **2015**, *10*, e0128553.
21. Malishev, R.; Arad, E.; Bhunia, S.K.; Shaham-Niv, S.; Kolusheva, S.; Gazit, E.; Jelinek, R. Chiral modulation of amyloid beta fibrillation and cytotoxicity by enantiomeric carbon dots. *Chem. Commun.* **2018**, *54*, 7762–7765. [CrossRef] [PubMed]
22. Zhou, Y.; Sun, Y.; Ma, Q.H.; Liu, Y. Alzheimer's disease: Amyloid-based pathogenesis and potential therapies. *Cell Stress.* **2018**, *2*, 150–161. [CrossRef] [PubMed]
23. Ravikirana, B.; Mahalakshmi, R. Unusual post-translational protein modifications: The benefits of sophistication. *RSC Adv.* **2014**, *4*, 33958–33974. [CrossRef]
24. Osna, N.A.; Carter, W.G.; Ganesan, M.; Kirpich, I.A.; McClain, C.J.; Petersen, D.R.; Shearn, C.T.; Tomasi, M.L.; Kharbanda, K.K. Aberrant post-translational protein modifications in the pathogenesis of alcohol-induced liver injury. *World J. Gastroenterol.* **2016**, *22*, 6192–6200. [CrossRef] [PubMed]
25. Grau, S.; Baldi, A.; Bussani, R.; Tian, X.; Stefanescu, R.; Przybylski, M.; Richards, P.; Jones, S.A.; Shridhar, V.; Tim Clausen, T.; et al. Implications of the serine protease HtrA1 in amyloid precursor protein processing. *Proc. Natl. Acad. Sci. USA* **2005**, *102*, 6021–6026. [CrossRef] [PubMed]
26. Gieldon, A.; Zurawa-Janicka, D.; Jarzab, M.; Wenta, T.; Golik, P.; Dubin, G.; Lipinska, B.; Ciarkowski, J. Distinct 3D architecture and dynamics of the human Htra2(Omi) protease and its mutated variants. *PLoS ONE* **2016**, *11*, e0161526. [CrossRef] [PubMed]
27. Goo, H.G.; Rhim, H.; Kang, S. Pathogenic role of serine protease HtrA2/Omi in neurodegenerative diseases. *Curr. Protein Pept. Sci.* **2017**, *18*, 746–757. [CrossRef]
28. Singh, M.K.; Sharma, B.; Tiwari, P.C. The small heat shock protein Hsp27: Present understanding and future prospects. *J. Therm. Biol.* **2017**, *69*, 149–154. [CrossRef]
29. Landry, J.; Lambert, H.; Zhou, M.; Lavoie, J.N.; Hickey, E.; Weber, L.A.; Anderson, C.W. Human HSP27 is phosphorylated at serines 78 and 82 by heat shock and mitogen-activated kinases that recognize the same amino acid motif as S6 kinase II. *J. Biol. Chem.* **1992**, *267*, 794–803.

30. Katsogiannou, M.; Andrieu, C.; Rocchi, P. Heat shock protein 27 phosphorylation state is associated with cancer progression. *Front. Genet.* **2014**. [CrossRef]
31. Lowenson, J.D.; Clarke, S.; Roher, A.E. Chemical modifications of deposited amyloid-β peptides. *Methods Enzym.* **1999**, *309*, 89–105.
32. Takahashi, O.; Kirikoshi, R.; Manabe, N. Racemization of serine residues catalyzed by dihydrogen phosphate Ion: A computational Study. *Catalysts* **2017**, *7*, 363. [CrossRef]
33. Shapira, R.; Austin, G.E.; Mirra, S.S. Neuritic plaque amyloid in Alzheimer's disease is highly racemized. *J. Neurochem.* **1988**, *50*, 69–74. [CrossRef] [PubMed]
34. Kaneko, I.; Morimoto, K.; Kubo, T. Drastic neuronal loss in vivo by beta-amyloid racemized at Ser (26) residue: Conversion of non-toxic [D-Ser (26)] beta-amyloid 1–40 to toxic and proteinase-resistant fragments. *Neuroscience* **2001**, *104*, 1003–1011. [CrossRef]
35. Tomiyama, T.; Asano, S.; Furiya, Y.; Shirasawa, T.; Endo, N.; Mori, H. Racemization of Asp23 residue affects the aggregation properties of Alzheimer amyloid beta protein analogues. *J. Biol. Chem.* **1994**, *269*, 10205–10208.
36. Shimizu, T.; Fukuda, H.; Murayama, S.; Izumiyama, N.; Shirasawa, T. Iso-aspartate formation at position 23 of amyloid beta peptide enhanced fibril formation and deposited onto senile plaques and vascular amyloids in Alzheimer's disease. *J. Neurosci. Res.* **2002**, *70*, 451–461. [CrossRef]
37. Warmack, R.A.; Boyer, D.R.; Zee, C.T.; Richards, L.R.; Sawaya, M.R.; Cascio, D.; Gonen, T.; Eisenberg, D.S.; Clarke, S.G. Structure of A-β (20–34) with Alzheimer's-associated isomerization at Asp23 reveals a distinct protofilament interface. *Nat. Commun.* **2019**, *10*, 3357. [CrossRef]
38. Kumar, S.; Singh, S.; Hinze, D.; Josten, M.; Sahl, H.G.; Siepmann, M.; Walter, J. Phosphorylation of amyloid-β peptide at serine-8 attenuates its clearance via insulin-degrading and angiotensin-converting enzymes. *J. Biol. Chem.* **2012**, *287*, 8641–8651. [CrossRef]
39. Barykin, E.P.; Petrushanko, I.Y.; Kozin, S.A.; Telegi, G.B.; Cherno, A.S.; Lopina, O.D.; Radko, S.P.; Mitkevich, V.A.; Makarov, A.M. Phosphorylation of the amyloid-beta peptide inhibits zinc-dependent aggregation, prevents Na,K-ATPase inhibition, and reduces cerebral plaque deposition. *Front. Mol. Neurosci.* **2018**, *11*, 302. [CrossRef]
40. Rezaei-Ghaleh, N.; Amininasab, M.; Kumar, S.; Walter, J.; Zweckstetterc, M. Phosphorylation modifies the molecular stability of β-amyloid deposits. *Nat. Commun.* **2016**, *7*, 11359. [CrossRef]
41. Milton, N.G. Phosphorylation of amyloid-beta at the serine 26 residue by human cdc2 kinase. *Neuroreport* **2001**, *12*, 3839–3844. [CrossRef] [PubMed]
42. Kumar, S.; Wirths, O.; Stüber, K.; Wunderlich, P.; Koch, P.; Theil, S.; Rezaei-Ghaleh, N.; Zweckstetter, M.; Bayer, T.A.; Brüstle, O.; et al. Phosphorylation of the amyloid β-peptide at Ser26 stabilizes oligomeric assembly and increases neurotoxicity. *Acta Neuropathol.* **2016**, *131*, 525–537. [CrossRef] [PubMed]
43. Zhao, J.; Wang, P.; Li, H.; Gao, Z. Nitration of Y10 in Aβ1–40: Is it a compensatory reaction against oxidative/nitrative stress and Aβ aggregation? *Chem. Res. Toxicol.* **2015**, *28*, 401–407. [CrossRef] [PubMed]
44. Ribó, J.M.; Hochberg, D. Concept Paper. Chemical basis of biological homochirality during the abiotic evolution stages on Earth. *Symmetry* **2019**, *11*, 814. [CrossRef]
45. Mori, H.; Ishii, K.; Tomiyama, T.; Furiya, Y.; Sahara, N.; Asano, S.; Endo, N.; Shirasawa, T.; Takio, K. Racemization: Its biological significance on neuropathogenesis of Alzheimer's disease. *Tohoku J. Exp. Med.* **1994**, *174*, 251–262. [CrossRef]
46. Kasim, J.K.; Kavianinia, I.; Harris, P.W.R.; Brimble, M.A. Three decades of amyloid beta synthesis: Challenges and advances. *Front. Chem.* **2019**, *7*, 472. [CrossRef]
47. Dyakin, V.V.; Lucas, J. Non-equilibrium phase transition in biochemical—Systems. Chain of chirality transfer as Determinant of Brain Functional Laterality. Relevance to Alzheimer disease and cognitive psychology. In Proceedings of the Alzheimer's Association International Conference (AAIC-2017), London, UK, 16–20 July 2017.
48. Ornes, S. Core concept: How nonequilibrium thermodynamics speaks to the mystery of life. *Proc. Natl. Acad. Sci. USA* **2017**, *114*, 423–424. [CrossRef]
49. Schrödinger, E. *What is Life? The Physical Aspect of the Living Cell*; Cambridge University Press: Cambridge, UK, 1994.

50. Nansheng, Z. The role of homochirality in evolution. In *Advances in BioChirality*; Zucchi, C., Caglioti, L., Palyi, G., Eds.; Elsevier Science: Amsterdam, The Netherlands, 1999.
51. Hsu, Y.H.; Chen, Y.-W.; Wu, M.-H.; Tu, L.H. Protein glycation by glyoxal promotes amyloid formation by Islet Amyloid polypeptide. *Biophys. J.* **2019**, *116*, 2304–2313. [CrossRef]
52. Zhang, A.; Qi, W.; Good, T.A.; Fernandez, E.J. Structural differences between Aβ (1–40) intermediate oligomers and fibrils elucidated by proteolytic fragmentation and hydrogen/deuterium exchange. *Biophys. J.* **2009**, *96*, 1091–1104.

Article

Spontaneous Chiral Symmetry Breaking and Entropy Production in a Closed System

Dilip Kondepudi [1,*] and Zachary Mundy [2]

1 Department of Chemistry, Wake Forest University, Winston-Salem, NC 27109, USA
2 BASF, 100 Park Avenue, Florham Park, NJ 07932, USA; zachary.mundy@basf.com
* Correspondence: dilip@wfu.edu

Received: 2 March 2020; Accepted: 14 April 2020; Published: 6 May 2020

Abstract: In this short article, we present a study of theoretical model of a photochemically driven, closed chemical system in which spontaneous chiral symmetry breaking occurs. By making all the steps in the reaction elementary reaction steps, we obtained the rate of entropy production in the system and studied its behavior below and above the transition point. Our results show that the transition is similar to a second-order phase transition with rate of entropy production taking the place of entropy and the radiation intensity taking the place of the critical parameter: the steady-state entropy production, when plotted against the incident radiation intensity, has a change in its slope at the critical point. Above the critical intensity, the slope decreases, showing that asymmetric states have lower entropy than the symmetric state.

Keywords: chiral symmetry breaking; entropy production; closed systems; nonequilibrium; dissipative structures

---

1. Introduction

Modern thermodynamics, formulated in the 20th century by Onsager [1], De Donder [2], Prigogine [3,4], and others, introduced a critical concept lacking in its classical formulation: rate of entropy change and its relationship to irreversible processes. Classical thermodynamics was concerned with functions of state, such as energy and entropy, and their change from one equilibrium state to another. Absent from this theory of states is consideration of the rates of processes. Changes in entropy for infinitely slow *reversible processes* are calculated using the relation dS = dQ/T, (in which dS is the change in entropy, T is the temperature in Kelvin, and dQ is the heat exchanged between a system and its exterior). However, for changes that take place in a finite time due to *irreversible processes,* the theory does not specify a way of calculating the entropy change; it is only stated that dS > dQ/T. Modern thermodynamics is a theory of processes in which thermodynamic forces and the flows they drive are identified and the rate of entropy production is expressed in terms of these thermodynamic forces and flows [5,6]. More specifically, the rate of entropy production per unit volume, σ, is expressed in terms of the forces and flows as

$$\sigma = \frac{ds}{dt} = \sum\nolimits_k F_k J_k \tag{1}$$

in which s is the entropy density, F_k are the thermodynamic forces and J_k are thermodynamic flows. A temperature gradient, for example, is the thermodynamic force, F_k, that drives thermodynamic flow, J_k, of heat current. The force that drives chemical reactions has been identified as *affinity* [2,5,6] and the corresponding flow is the rate of conversion form reactants to products. This flow is expressed as the time derivative dξ/dt (mol/s) of the extent of reaction ξ [5,6]. For an elementary chemical reaction step, the rate of entropy production can be written in terms of the forward reaction rate, R_f, and the reverse reaction rate, R_r [6]

$$\sigma = R\left(R_f - R_r\right)\ln\left(\frac{R_f}{R_r}\right) \quad (2)$$

in which R is the gas constant. The total entropy production for a sequence of reactions is the sum of entropy productions of each reaction (indexed by k) [6]

$$\sigma = R\sum_k \left(R_{kf} - R_{kr}\right)\ln\left(\frac{R_{kf}}{R_{kr}}\right) \quad (3)$$

We note that calculations of entropy production using the above formulas require that the reverse reaction rates, R_r, are non-zero. In considering kinetic equations of a chemical system, often the reverse rates have high reaction barriers and, correspondingly, very low reaction rates, and are assumed to be "zero" because they are negligible compared to the forward rates. Low reverse reaction rates keep the system from evolving to equilibrium state. In the model we will present below, the system is driven far from equilibrium by an inflow of radiation. In the absence of radiation, our system evolves to equilibrium—there are no very high reaction barriers to keep it from reaching its equilibrium state.

It is well known that nonequilibrium chemical systems can undergo spontaneous symmetry-breaking transitions to organized structures called *dissipative structures* [6,7]. The large class of dissipative structures includes spatial patterns, chemical clocks, and structures with chiral asymmetry. These structures arise when a nonequilibrium system becomes unstable and undergoes a transition to new organized state. The model we present below shows a chiral symmetry breaking transition to an asymmetric state as the intensity of radiation increases.

The study of spontaneous chiral symmetry breaking in chemical systems has a long history, starting in the 1950s with the first model by Frank [8]. In the 1970s, when the theory of dissipative structures was developed, instability, bifurcation, and spontaneous symmetry breaking in non-equilibrium systems became the foundation for the study of chiral asymmetry we see in nature [4,7,9–11]. A general theory of chiral symmetry breaking in chemical systems, independent of the details of the chemical kinetics, was formulated, and it was used to study the sensitivity of chiral symmetry breaking systems to small chiral influences [11–13]. Using this theory, it was possible to calculate the time scales needed for a chiral-symmetry-breaking chemical system to be influenced by the chiral asymmetry of electro-weak interaction in molecules [14], and it was found that this timescale is of the order to 10^4 years [15]. These results show that a lot of interesting and important general conclusions can be arrived at through using theoretical models [16,17]. Along these lines, we investigate the thermodynamic aspects of systems that spontaneously break chiral symmetry, using a model presented below.

Using nonequilibrium thermodynamics, we analyze the behavior of entropy production, σ, for a reaction scheme that consists of a photochemically driven *closed system* (that has no flow of matter). The incident radiation drives a generation–decomposition cycle of chiral molecules. At the critical intensity (above which the system breaks chiral symmetry), the slope of the rate of entropy production, σ, changes just as entropy does in a second-order phase transition. We have reported a similar result in a flow system with an inflow of reactants and outflow of products [18]. There is a basic difference, however: in our previous study, the slope of σ increased, but in our current study we find that the slope decreased, although in both cases the behavior of σ is similar to that in a second-order phase transition.

2. Materials and Methods

All reactions in our scheme are reversible so that the system can reach a state of chemical equilibrium. At equilibrium, the rate of entropy production is zero. The reaction scheme of this photochemically driven closed system is shown in Table 1. It is assumed to take place in a homogeneous aqueous phase in which radiation is incident. For a photochemical reaction, the intensity of a narrow band of wavelength is the relevant intensity, therefore it will have a low value compared to a typical intensity of blackbody radiation that includes all wavelengths—for example, radiation from the sun. An achiral molecule, T, is photochemically activated to an excited species or a more reactive isomer, TE, as shown in reaction (R1); II is the intensity of radiation of the exciting wavelength. TE can radiate its excitation energy and return to the unexcited state, T (R1a), through various processes. During the transition of TE to T, the emitted radiation undergoes scattering and thermalizes to the temperature of the system. (R1a) represents all the processes that keep TE and T at equilibrium (including thermal radiation [6]) in the absence of external radiation. In writing the reaction rates for T and TE, we only need to include additional term in the forward reaction that includes II and the all reverse reactions, TE to T, combined into one reaction rate. The excited species TE can also react with an achiral molecule S to form a chiral species, X_L or X_D (R2 and R3). The set of reactions R4a and R4b are elementary steps of an autocatalytic reaction for X_L with the intermediate species S_L; similar reaction steps result in the autocatalytic production of X_D, as shown in R5a and R5b. Reaction steps R6, R7 and R8 show reactions through which the species decompose into starting material S and T. The scheme explicitly has all the steps needed for the system to reach chemical equilibrium in the absence of radiation.

Table 1. A model photochemically driven reaction scheme in a closed system. The table lists all the elementary steps in the model. The excitation of T to the state TE drives the reaction that generate chiral species X_L and X_D. Reactions R4a, R4b, R5a, R5b are the autocatalytic steps for X_L and X_D. Autocatalysis and reaction R6 result in spontaneous breaking of chiral symmetry when the intensity of radiation, II, is above a critical intensity, II_C.

Chemical Reactions	Number
$S + II \rightleftharpoons TE$	(R1)
$T \rightleftharpoons TE$	(R1a)
$S + TE \rightleftharpoons X_L$	(R2)
$S + TE \rightleftharpoons X_D$	(R3)
$S + X_L \rightleftharpoons S_L$	(R4a)
$S_L + TE \rightleftharpoons 2X_L$	(R4b)
$S + X_D \rightleftharpoons S_D$	(R5a)
$S_D + TE \rightleftharpoons 2X_D$	(R5b)
$X_L + X_D \rightleftharpoons P + W$	(R6)
$P \rightleftharpoons 2S$	(R7)
$W \rightleftharpoons 2T$	(R8)

The above model is a variation of the models in our earlier studies [11,12,15] which are modifications of the original model of Frank [8]. The modifications allow us to analyze non-equilibrium symmetry breaking and rate of entropy production. Models such as this are used to extract general properties that are not model dependent. Examples of such properties are the qualitative behavior of steady-state rate of entropy production as a function of a parameter that drives the system away from equilibrium (such as the incident radiation intensity II in the above model). The difference in concentration between enantiomers of a chiral species as a function of a parameter, such as the intensity II, is generally parabolic, as predicted by bifurcation theory based on the symmetry group (mirror symmetry in this case) of the system, regardless of the details of the chemical reactions that break chiral symmetry.

Though there is currently no known reaction that has all the properties in the above model, the reaction has no steps that are implausible. For example, reactions (R1), (R1a), (R2) and (R3) comprise a photoaddition reaction that produces a chiral compound. An example is the following reaction series [19,20]:

$$(Ph)_2\, C = CH(R^1) + h\nu \rightarrow [(Ph)_2\, C = CH(R^1)]^* \quad \text{(R9)}$$
$$[(Ph)_2\, C = CH(R^1)]^* + (R^2)OH \rightarrow [(Ph)_2\, HC - CH(R^1)(OR^2)]_L \quad \text{(R10)}$$
$$[(Ph)_2\, C = CH(R^1)]^* + (R^2)OH \rightarrow [(Ph)_2\, HC - CH(R^1)(OR^2)]_D \quad \text{(R11)}$$

in which Ph is the phenyl group and $R^1 = CH_3$ or C_2H_5 and $R^2 = CH_3, C_2H_5$ or C_3H_7. In the reaction (R9), a photon is absorbed by the electrons in the C=C double bond and the molecule transitions to a reactive excited state $[(Ph)_2 C = CH(R^1)]^*$. In the addition reaction shown in (R10) and (R11), the excited molecule reacts with an alcohol, (R^2) OH, and produces a chiral compound $(Ph)_2$ HC − **C**$H(R^1)(OR^2)$ in the L and D enantiomeric forms. In this compound, the carbon shown in boldface is a chiral carbon (its tetrahedral bonds to four different groups makes it so). Other examples of photoaddition reactions producing chiral products from achiral reactants can be found in [19]. We note that TE need not be an excited state; it could be a different, more reactive isomer of the T [19].

The reactions (R4a)–(R5b) are steps leading to chiral autocatalysis. This involves the formation of a chiral complex of S and X in their enantiomeric forms. Examples of chiral complexes resulting in reactions with a high degree of chiral selectivity have been known for a long time [21,22]. In an article published in 1984, we noted some mechanisms that are based on chiral ligands in rhodium phosphine catalysts which could lead to chiral autocatalysis [12,22]. To date, there are several chirally autocatalytic reactions that have been experimentally studied. Chiral symmetry breaking was noticed and systematically studied first in NaClO3 in 1990 [23], and in 1995 chiral autocatalysis and amplification of small initial enantiomeric excess was reported in inorganic reactions involving cobalt complexes [24] and in organic reactions involving alkylation of aldehydes [25]. Since then, these and closely related systems have been extensively experimentally studied and the mechanisms of chiral autocatalysis have been investigated [26–31]. A variant of chiral symmetry breaking in stirred crystallization was reported in 2005, and it too has been studied extensively [32,33]. The mechanisms of chiral autocatalysis vary in these systems: in crystallization, it is secondary nucleation, in the organic and inorganic reactions, cluster/complex formation seems to be involved. Reactions (R4a) and (R5a) may be thought of as a simple form of chiral complex formation. As was noted in a review [27], the exact details of chiral catalysis are not of significance for the general properties symmetry-breaking bifurcation and thermodynamic properties of such systems. In particular, properties such as phase-transitions-like behavior we present here are quite independent of the details of chemical kinetics. Examples of reaction (R6), the dimer formation of enantiomers, are also known; in fact, such dimerization of certain chiral catalysts leads to asymmetric amplification [34].

For the above theoretical model (R1)–(R8), the corresponding forward and reverse rate for each reaction are written as follows, in which concentrations are shown explicitly as functions of time:

$$\mathrm{R1f = (k1f + k1\ II)T[t],\ R1r = k1r\ TE[t]} \tag{4}$$

$$\mathrm{R2f = k2f\ S[t]\ TE[t],\ R2r = k2r\ X_L[t]} \tag{5}$$

$$\mathrm{R3f = k3f\ S[t]\ TE[t],\ R3r = k3r\ X_D[t]} \tag{6}$$

$$\mathrm{R4af = k4af\ S[t]\ X_L[t],\ R4ar = k4ar\ S_L[t]} \tag{7}$$

$$\mathrm{R4bf = k4bf\ S_L[t]\ TE[t],\ R4br = k4br\ (X_L[t])^2} \tag{8}$$

$$\mathrm{R5af = k5af\ S[t]\ X_D[t],\ R5ar = k5ar\ S_D[t]} \tag{9}$$

$$\mathrm{R5bf = k5bf\ S_D[t]\ TE[t],\ R5br = k5br\ (X_L[t])^2} \tag{10}$$

$$\mathrm{R6f = k6f\ X_L[t]X_D[t],\ R6r = k6r\ P[t]W[t]} \tag{11}$$

$$\mathrm{R7f = k7f\ P[t],\ R7r = k7r\ (S[t])^2} \tag{12}$$

$$\mathrm{R8f = k8f\ W[t],\ R8r = k8r\ (T[t])^2} \tag{13}$$

In these equations, the rate constants are written as k1f, k1r etc., and the concentration are written as T[t], S[t], etc. In terms of these forward and reverse rates, the rate equations for the concentrations can be written as:

$$\mathrm{d\ T[t]/dt = -R1f + R1r + 2R8f - 2R8r} \tag{14}$$

$$\mathrm{d\ TE[t]/dt = -R1f + R1r - R2f + R2r - R3f + R3r - R4bf + R4br - R5bf + R5br} \tag{15}$$

$$d\,S[t]/dt = -R2f + R2r - R3f + R3r - R4af + R4ar - R5af + R5ar + 2R7f - 2R7r \quad (16)$$

$$d\,S_L[t]/dt = R4af - R4ar - R4bf + R4br \quad (17)$$

$$d\,S_D[t]/dt = R5af - R5ar - R5bf + R5br \quad (18)$$

$$d\,X_L[t]/dt = R2f - R2r - R4af + R4ar + 2R4bf - 2R4br - R6f + \quad (19)$$

$$d\,X_D[t]/dt = R3f - R3r - R5af + R5ar + 2R5bf - 2R5br - R6f + R6r \quad (20)$$

$$d\,P[t]/dt = R6f - R6r - R7f + R7r \quad (21)$$

$$d\,W[t]/dt = R6f - R6r - R8f + R8r \quad (22)$$

This set of coupled non-linear equations were solved numerically using Mathematica NDSolve. NDSolve is a Mathematica command that has the following structure: *NDSolve[{Equations}, {y_i},{t, t_{min}, t_{max}}]*, in which "Equations" are the differentials equations for the set of functions {y_i} with t as the independent variable; numerical solutions are obtained in the range t_{min}, to t_{max}. More details can be found in the online documentation that comes with Mathematica. The rate constants used for the numerical solutions are summarized in Figure 1. In assigning values to rate constants, there are consistency conditions that must be met. For example, since reactions (R4a) and (R4b) together are equivalent to (R2), the products of the equilibrium constants of R4a and R4b must equal the equilibrium constant of R2. This gives us the following condition for the rate constants:

$$(k4af/k4ar)(k4bf/k4br) = k2f/k2r \quad (23)$$

$$k1f = (10^{-3})\,k1r$$

$$k1r = 1.0$$

$$kI = 1$$

$$k2f = 0.5$$

$$k2r = 0.001$$

$$k3f = k2f$$

$$k3r = k2r$$

$$k4af = 10^6$$

$$k4ar = 10^5$$

$$k4bf = 10^3$$

$$k4br = k4bf\left(\frac{k4af}{k4ar}\right)\left(\frac{k2r}{k2f}\right)$$

$$k5af = k4af$$

$$k5ar = k4ar$$

$$k5bf = k4bf$$

$$k5br = k4br$$

$$k6f = 100.0$$

$$k6r = k6f\left(\frac{k1f}{k1r}\right)^2\left(\frac{k2f}{k2r}\right)\left(\frac{k3f}{k3r}\right)\left(\frac{k7f}{k7r}\right)\left(\frac{k8f}{k8r}\right)$$

$$k7f = 0.5$$

$$k7r = 5.0$$

$$k8f = k7f$$

$$k8r = k7r$$

$$II = variable$$

Figure 1. The figure shows numerical values assigned to rate constants to obtain numerical solution for the rate equations of the system. Units of volume are assumed to be liters (L) and concentrations mol/L (M). Units of variable parameter II may be thought of as W/m^2. If II is thought of as a radiation from the sun, its blackbody temperature is very high compared to the temperature of the system. All rate constants are assumed to have the appropriate units, though not written explicitly.

Numerical values were assigned to rate constants so as to fulfill these requirements. The units were chosen so that all concentrations are in mol/L. Assigned numerical values are such that the concentrations of the reactants have realistic values. Symmetric (racemic) and asymmetric states are parametrized by $\alpha = (X_L - X_D)$, in the symmetric state $\alpha = 0$ and in the asymmetric state $\alpha \neq 0$.

3. Results

The rate equations were first solved for an equilibrium state where the incident radiation intensity, II, was set to 0. The initial concentrations of species S and T were set to 0.01 M and the initial concentrations of all other species were set to 0.0 M. Under these conditions, the system evolves to its racemic equilibrium state, in which $\alpha = 0$.

The simulation code was run for sufficient time (about 10^4 s) to ensure the concentrations of all species have reached a steady state, which is the equilibrium state. At t = 10^4 s, the concentrations at equilibrium were: S = T = 8.478×10^{-3} M, TE = 8.477×10^{-6} M, $S_L = S_D = 3.047 \times 10^{-6}$ M, $X_L = X_D = 3.593 \times 10^{-5}$ M, and P = W = 7.188×10^{-4} M. The conversion of the initial species S and T compared to other species was rather small for the numerical values of the rate constants shown in Figure 1. By choosing a different set of rate constants, the conversion could be increased. The numerical values confirm that complete symmetry of the system was maintained when no incident radiation is present. Figure 2 shows the time evolution of the chiral species S_L, S_D, X_L, X_D from t = 0 s, to t = 1000 s. As the system evolved to its equilibrium state, the entropy production σ was monitored; initially, it took a nonzero value but, as expected, its value decreased to zero at the equilibrium state.

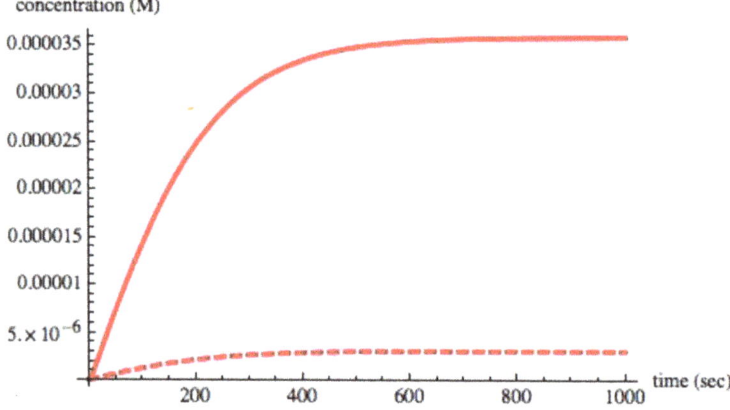

Figure 2. Time evolution of chiral species to their equilibrium state in the model reaction scheme when II = 0 and S = T = 0.01 M at t = 0. Concentrations of all other species was set to zero at t = 0. Due to the symmetry of the system, at equilibrium, $S_L = S_D$ (dashed line) and $X_L = X_D$ (solid line), so the curves for the two enantiomers overlap.

We then used these values for a symmetric equilibrium state as initial values for a system subject to a radiation input, i.e., II > 0. This radiation input serves as a means to push the system away from thermodynamic equilibrium. A very small excess, about 0.1% (3×10^{-8} M) of X_L was introduced into the system as a "random fluctuation". If the system has the mechanism to break chiral symmetry, it will have a critical value II_C. At values of II < II_C, this excess 0.1% of X_L will decrease and the system will again evolve into a steady state where $X_L = X_D$; at values of II > II_C, the excess will increase and lead to a steady state in which $X_L > X_D$. In a real system, this slight perturbation may be due to a random fluctuation such as a local excess of one enantiomer that may then be amplified, resulting in a state of broken symmetry. The overall behavior of the system is summarized in Figure 3.

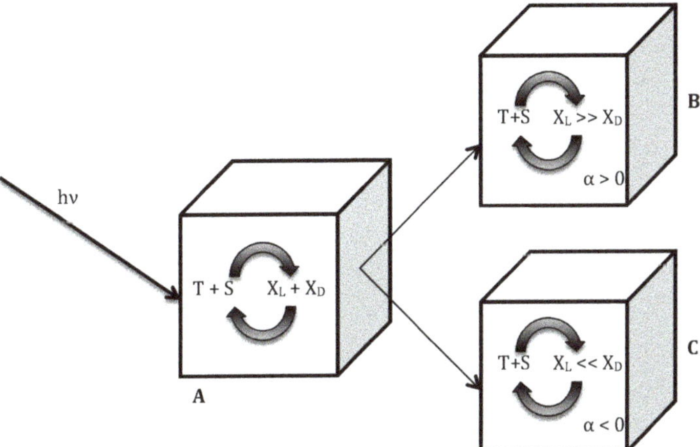

Figure 3. Schematic of reaction system. Radiation (shown as hν and denoted as II in the reaction scheme)) is incident on a closed chemical system. The radiation drives a generation–decomposition cycle of enantiomeric species X and other compounds, as shown in A. When II > II_C, the system evolves to one of two asymmetric states, B or C. In state B, the amount of $X_L > X_D$, and in state C, $X_D > X_L$. The asymmetry is parametrized by $\alpha = (X_L - X_D)$.

It was found that this system indeed has the mechanism needed for breaking chiral symmetry. At values of II < 0.004, the system evolved to a symmetric state corresponding to $X_L = X_D$ and $\alpha = 0$. Figure 4 shows the time evolution of the chiral species when II = 0.0030. A steady state is reached in approximately 600 s.

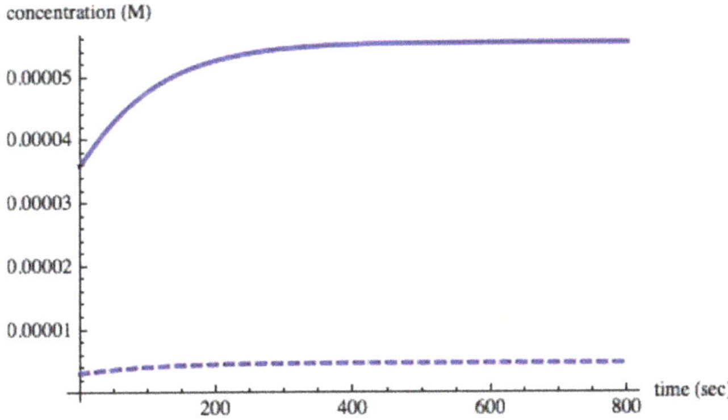

Figure 4. Time evolution of chiral species using equilibrium values for initial conditions when II = 0.003. The solid line represents X_L and X_D and the dashed S_L and S_D. Since symmetry is not broken, the amounts of each chiral species are exactly equal, hence the overlapping curves ($\alpha = 0$).

For values of II > 0.004, the small excess of 0.1% of XL in the initial concentration increased, thus driving all chiral species to an asymmetric state. A time evolution of the chiral species in an asymmetric state when II = 0.008 is shown in Figure 5.

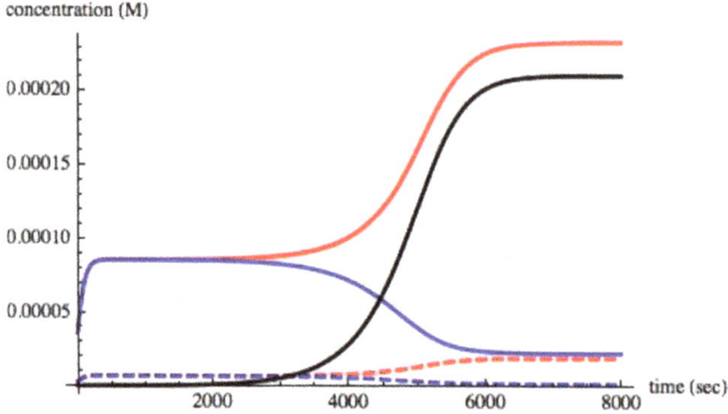

Figure 5. Time evolution of chiral species in an asymmetric state. Here, II = 0.008 and an asymmetric steady state is reached in about 7000 s. The blue solid and dashed lines represent X_D and S_D, respectively, and the red solid and dashed lines represent X_L and S_L, respectively. The black line represents α, which takes on a nonzero value once symmetry is broken.

As shown, steady state is reached about 7000 s. The time taken to reach steady state, the *relaxation time*, depends on the value of II and on the amount of initial excess (0.1% of X_L). As is well known in the study of stability of steady states [6,7,9], the initial exponential growth of the small enantiomeric excess depends on the eigenvalue of the unstable mode of the linearized equations derived from the set (14)–(22) around the initial state. These linearized equations are obtained by assuming a small perturbation of the concentrations, δC_k, from the initial steady state. This leads to a set of linear equations of the type $d\delta C_k/dt = \Sigma_l L_{kl}\delta C_l$ [6,7,9]. The eigenvalues of L_{kl} with positive real parts are the exponents that determine the initial rate of growth of the enantiomeric excess. However, the later growth and leveling off at the steady state depend on the kinetics and rate constants. In general, near the critical point II_C, the relaxation time is long, the so-called "critical slowing", but as the value of II increases, the growth rate becomes faster and the relaxation time decreases.

To determine the relation between the steady-state value of α on II, the reaction was run for various values of II, from 0.0035 to 0.0045, and the corresponding steady-state values of α were obtained. As noted above, the time it takes for the system to reach an asymmetric steady state depends on the value of II; close to the critical value II_C (0.004 in this case), the relaxation to asymmetric steady state is slower and it becomes faster as the value of II increases. The exact quantitative relationship between relaxation time and II depends on the kinetics and rate constants and not of significance to the current study. In these runs, to obtain both positive and negative branches of α, the initial condition with a 0.1% excess of X_D was also included. The dependence of α on II is shown in Figure 6, demonstrating the typical bifurcation of asymmetric states above the *critical value* $II_C = 0.004$. As is expected, in a chiral symmetry breaking transition, the values of α above the critical point are parabolic.

With these results, we now turn to the rate of entropy production σ. As stated above, initially the system is in the state of equilibrium, with the intensity of radiation II = 0 and $\sigma = 0$. Then the intensity of the radiation II is increased to a non-zero value and the rate of entropy production σ is monitored. Initially, σ sharply increases and, as the system reaches its steady-state, corresponding to the set value of II, the entropy production also reaches its steady state value. Figure 7 shows the evolution of σ from its value when t = 0, to its final steady-state value when II = 0.008. The final steady state value of σ is small compared to its initial value, but it is nonzero.

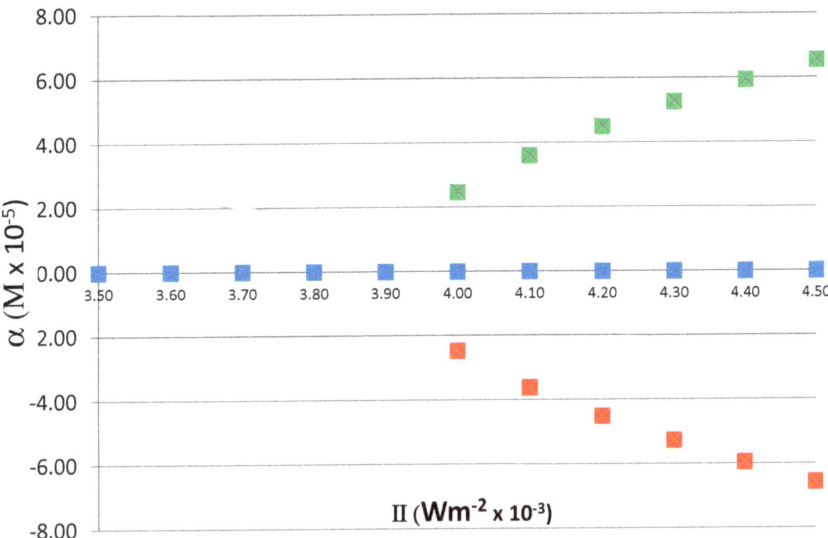

Figure 6. Dependence of α on II. Units of α are M and II are Wm^{-2}. Steady-state values of α are plotted as a function of II. When II > II_C, α takes a positive (X_L > X_D, shown in green X) or negative (X_L < X_D, shown in red X) value, depending on the random perturbation that drives the system away from the unstable racemic state α = 0. The blue Xs show the symmetric branch which is unstable above the critical point. In the region II > II_C, α increases in a characteristically parabolic manner.

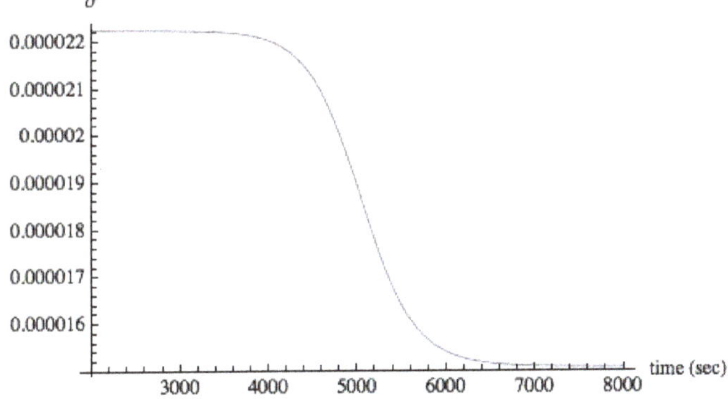

Figure 7. Rate of entropy production in an asymmetric state when II = 0.008. Units of σ are $JK^{-1}L^{-1}s^{-1}$. As in Figure 3, steady state is reached at about 7000 s. Although it may appear that σ is approaching 0, it is not so; σ maintains a nonzero value at steady state after symmetry is broken.

We would like to note that the results shown above do not depend crucially on the particular values assigned to rate constants. Whether symmetry breaking occurs or does not depends mostly on the mechanism of reactions in the model, not on a narrow range of values of rate constants. In general, qualitative properties change drastically due to small changes in rate constants only in singular cases. The presented model is not singular. In our study, we have tried a range of rate constants and observed symmetry breaking. A typical value is presented in this article.

Our objective is to study the behavior of σ as the intensity of radiation II moves form a value below to a value above the critical value II_C. In a previous study [18] of an open system, σ behaved as entropy does in a second order phase transition: its slope is discontinuous at the transition point.

In addition, we noted that its slope increased above the critical point. This behavior was consistent with the Maximum Entropy Production (MEP) hypothesis [35–40] that states that non-equilibrium steady states maximize the rate of entropy production. In other words, the breaking of symmetry was a reflection of the general tendency of non-equilibrium systems and, from this point of view, it was only to be expected. This would imply that biochemical asymmetry is a consequence of MEP. To date, there is no general proof for MEP; indeed, in our own investigation we found that MEP was valid for some systems but not all. MEP, in general, is a controversial hypothesis and its applications to various complex systems have been questioned [41]. Hence, we investigated MEP in the context of chiral symmetry breaking.

Figure 8 shows the behavior of entropy production in the closed system we present here. It shows a change in the slope at the transition point, IIC, as in our previous study. From the point of view of symmetry breaking transition, we see that entropy production behaves in a way that is similar to that of entropy in a second-order phase transition. However, unlike the previous result, the slope decreases above the critical point. As indicated in Figure 8, when $II > II_C$, if initially we set the system in a non-equilibrium symmetric state, which is an extrapolation of the symmetric state below II_C, the system evolves to an asymmetric state in which σ is lower. The extrapolation of the symmetric state beyond the critical value II_C is possible on a computer because, without a small perturbation in X_L or X_D, or other chiral species, the system stays in an unstable symmetric steady state. With a small perturbation, it evolves to the stable asymmetric state. The time evolution of the system from this initial state to its final asymmetric steady state results in the decrease in σ, thus indicating that the stable asymmetric state is associated with a lower value of σ compared with that of a symmetric state. Thus, we see that the entropy production in a closed system is not consistent with MEP, because MEP would predict higher values for σ in the asymmetric state.

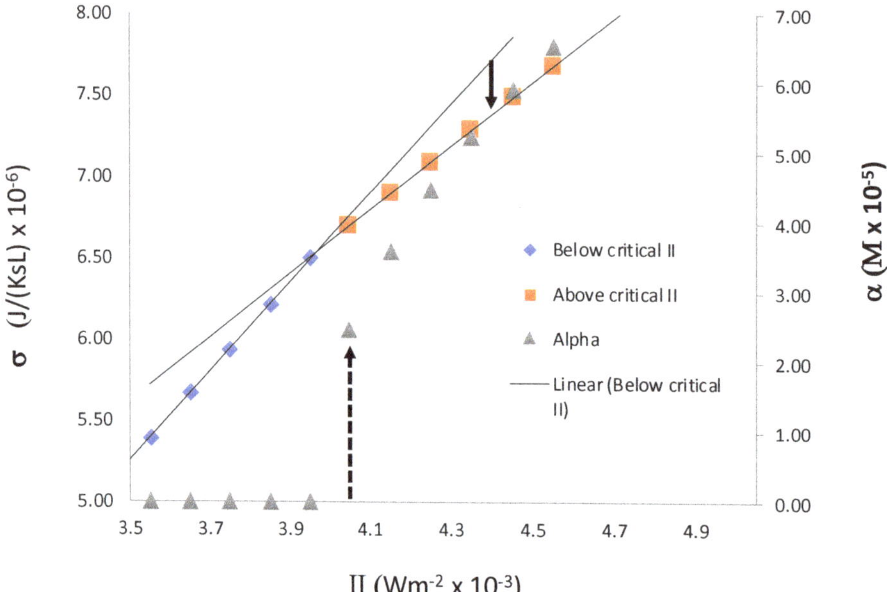

Figure 8. Rate of entropy production σ and α as a function of II. The critical value $II_C = 0.0040$. Beyond this point, α takes on nonzero values and there is a decrease in the slope of σ. Units of II are Wm^{-2}. The dashed arrow indicates a transition from an unstable state, where $\alpha = 0$, to a stable asymmetric state where α is nonzero. In the region $II > II_C$ rate of entropy production of asymmetric state is lower than that of the symmetric state. The solid arrow shows that transition from an unstable symmetric state to a stable asymmetric state results in the lowering of σ, the rate of entropy production.

4. Concluding Remarks

Our results show several aspects of chiral symmetry breaking. First, they show that Frank's original model can be modified with several additional steps—all of which are possible elementary chemical reaction steps—to demonstrate that spontaneous chiral symmetry breaking can occur in a photochemically driven closed system. Our model is motivated by the fact that the earth is a essentially a closed system (except for a very small influx in interstellar matter such as meteorites) and the evolution of life was driven by incident solar radiation. The incident radiation drives a cycle of generation and decomposition of chiral compounds that, at a sufficiently high intensity of radiation, makes a transition to a state of broken chiral symmetry. This example demonstrates that life on earth could have evolved under such conditions of prebiotic molecular chiral asymmetry.

From a thermodynamic viewpoint, this study also confirms that chiral symmetry breaking transitions are similar to second-order phase transitions, with entropy production taking the place of entropy. We note that the general qualitative features of bifurcation of asymmetric states and change in the behavior entropy production at the critical point, shown in Figures 6 and 8, are a consequence of the two-fold symmetry of the system, not the particularities of the chemical reactions in the model. In fact, using the system's symmetry group, it is possible to derive the following generic equation for the evolution of α near the critical point: $d\alpha/dt = -A\alpha^3 + B\alpha + C$, in which A, B and C are functions of the kinetic constants of the chiral symmetry breaking chemical reactions [11–13,15]. Finally, we show that the behavior of entropy production in this chiral-symmetry-breaking system is not consistent with the MEP hypothesis, because MEP implies that the state of broken symmetry will have a higher rate of entropy production compared to a symmetric state, but we find the opposite to be the case in this model. In a system with an inflow of reactants and outflow of products, however, the entropy production was higher in the asymmetric state. This indicates that MEP is valid for a certain class of systems, but what this class is has not yet been clearly identified.

Author Contributions: Conceptualization, D.K.; methodology, D.K. and Z.M.; software and Mathematica Code, D.K. & Z.M. All authors have read and agreed to the published version of the manuscript.

Funding: Wake Forest University, Winston Salem, NC 27109, USA.

Conflicts of Interest: The authors declare no conflict of interest.

References

1. Onsager, L. Reciprocal Relations in Irreversible Processes. *Phys. Rev.* **1931**, *37*, 405–426. [CrossRef]
2. De Donder, T.; Van Rysselberghe, P. *Affinity*; Stanford University Press: Menlo Park, CA, USA, 1936.
3. Prigogine, I. *Etude Thermodynamique des Phénomènes Irréversibles*; Editions Desoer: Liège, Belgium, 1947.
4. Prigogine, I. *Introduction to Thermodynamics of Irreversible Processes*; John Wiley: London, UK, 1967.
5. De Groot, S.R.; Mazur, P. *Non-Equilibrium Thermodynamics*; North Holland Press: Amsterdam, The Netherlands, 1969.
6. Kondepudi, D.K.; Prigogine, I. *Modern Thermodynamics: From Heat Engines to Dissipative Structures*, 2nd ed.; John Wiley: New York, NY, USA, 2015.
7. Nicolis, G.; Prigogine, I. *Self-Organization in Non-equilibrium Systems*; Wiley: New York, NY, USA, 1977.
8. Frank, F.C. Spontaneous asymmetric synthesis. *Biochem. Biophys. Acta* **1953**, *11*, 459–463. [CrossRef]
9. Haken, H. *Synergetics: An Introduction*; Springer: Berlin, Germany, 1977.
10. Kondepudi, D.K.; Hegstrom, R.A. The Handedness of the Universe. *Sci. Am.* **1990**, *262*, 108–115.
11. Kondepudi, D.K.; Nelson, G.W. Chiral Symmetry Breaking in Nonequilibrium Systems. *Phys. Rev. Lett.* **1983**, *50*, 1023–1026. [CrossRef]
12. Kondepudi, D.K.; Nelson, G. Chiral Symmetry Breaking states and Their Sensitivity in Nonequilibrium Systems. *Physica A* **1984**, *125*, 465–496. [CrossRef]
13. Kondepudi, D.K. State Selection Dynamics in Symmetry Breaking Transitions. In *Noise in Nonlinear Dynamical Systems*; Moss, F., McClintock, P.V., Eds.; Cambridge University Press: Cambridge, UK, 1989; Volume 2, pp. 251–270.

14. Hegstrom, R.A.; Rein, D.W.; Sanders, P.G.H. Calculation of the parity nonconserving energy difference between mirror-image molecules. *J. Chem. Phys.* **1980**, *73*, 2329–2341. [CrossRef]
15. Kondepudi, D.K.; Nelson, G.W. Weak Neutral Currents and the Origin of Biomolecular Chirality. *Nature* **1985**, *314*, 438–441. [CrossRef]
16. Decker, P. The origin of molecular asymmetry through the amplification of "stochastic information" (noise) in bioids, open systems which can exist in several steady states. *J. Mol. Evol.* **1974**, *4*, 49–65. [CrossRef]
17. Plasson, R.; Kondepudi, D.K.; Bersini, H.; Commeyras, A.; Asakura, K. Emergence of homochirality in far-for-equilibrium systems: Mechanisms and role in prebiotic chemistry. *Chirality* **2007**, *19*, 589–600. [CrossRef]
18. Kondepudi, D.K.; Kapcha, L. Entropy production in chiral symmetry breaking transitions. *Chirality* **2008**, *20*, 524–528. [CrossRef]
19. Inoue, Y. Enantiodifferentiating Photosensitized Reactions. In *Chiral Photochemistry*; CRC Press: Boca Raton, FL, USA, 2004; Chapter 4.
20. Inoue, Y.; Ramamurthy, V. (Eds.) *Chiral Photochemistry*; Marcel Decker: New York, NY, USA, 2004.
21. Kagan, H.B. Chiral Ligands for Asymmetric Catalysis. In *Asymmetric Synthesis: Chiral Catalysis*; Morrison, J.D., Ed.; Academic Press: New York, NY, USA, 1985; Chapter 1; Volume 5, pp. 1–39.
22. Fryzuk, M.D.; Bosnich, B. Asymmetric synthesis. An asymmetric homogeneous hydrogenation catalyst which breeds its own chirality. *J. Am. Chem. Soc.* **1978**, *100*, 5491–5494. [CrossRef]
23. Kondepudi, D.K.; Kaufmann, R.; Singh, N. Chiral Symmetry Breaking in Sodium Chlorate Crystallization. *Science* **1990**, *250*, 975–976. [CrossRef] [PubMed]
24. Asakura, K.; Kobayashi, K.; Mizusawa, Y.; Ozawa, T.; Osanai, S.; Yoshikawa, S. Generation of an optically active octahedral cobalt complex by a chiral autocatalysis. *Physica D* **1995**, *84*, 72–78. [CrossRef]
25. Soai, K.; Shibata, T.; Morioka, H.; Choji, K. Asymmetric autocatalysis and amplification of enantiomeric excess of a chiral molecule. *Nature* **1995**, *378*, 767–768. [CrossRef]
26. Kondepudi, D.K.; Asakura, K.; Martin, R. Mechanism of Chiral Asymmetry Generation by Chiral Autocatalysis in the Preparation of Chiral Octahedral Cobalt Complex. *Chirality* **1998**, *10*, 343–348.
27. Kondepudi, D.K.; Asakura, K. Chiral Autocatalysis, Spontaneous Symmetry Breaking, and Stochastic Behavior. *Acc. Chem. Res.* **2001**, *34*, 946–954. [CrossRef]
28. Soai, K.; Shibata, T.; Sato, I. Enantioselective Automultiplication of Chiral Molecules by Asymmetric Autocatalysis. *Acc. Chem. Res.* **2000**, *33*, 382–390. [CrossRef]
29. Gehring, T.; Busch, M.; Schlageter, M.; Weingand, D. A concise summary of experimental facts about the Soai reaction. *Chirality* **2010**, *22* (Suppl. 1), E173–E182. [CrossRef]
30. Soai, K.; Kawasaki, T.; Matsumoto, A. Asymmetric Autocatalysis of Pyrimidyl Alkanol and Its Application to the Study on the Origin of Homochirality. *Acc. Chem. Res.* **2014**, *47*, 3643–3654. [CrossRef]
31. Blackmond, D.G. Autocatalytic Models for the Origin of Biological Homochirality. *Chem. Rev.* **2019**. [CrossRef]
32. Viedma, C. Chiral Symmetry Breaking During Crystallization: Complete Chiral Purity Induced by Nonlinear Autocatalysis and Recycling. *Phys. Rev. Lett.* **2005**, *94*, 065504–065506. [CrossRef] [PubMed]
33. Sögütoglu, L.C.; Steendam, R.R.; Meekes, H.; Vlieg, E.; Rutjes, F.P. Viedma ripening: A reliable crystallisation method to reach single chirality. *Chem. Soc. Rev.* **2015**, *44*, 6723–6732. [CrossRef] [PubMed]
34. Girard, C.; Kagan, H. Nonlinear Effects in Asymmetric Synthesis and Stereoselective Reactions: Ten Years of Investigation. *Angew. Chem. Int. Ed.* **1998**, *37*, 2922–2959. [CrossRef]
35. Swenson, R. Emergent attractors and the law of maximum entropy production: Foundations to a theory of general evolution. *Syst. Res.* **1989**, *6*, 187–197. [CrossRef]
36. Swenson, R.; Turvey, M.T. Thermodynamic Reasons for Perception—Action Cycles. *Ecol. Psychol.* **1991**, *3*, 317–348. [CrossRef]
37. Martyushev, L.M.; Seleznev, V.D. Maximum entropy production principle in physics, chemistry and biology. *Phys. Rep.* **2006**, *426*, 1–45. [CrossRef]
38. Maysman, F.J.R.; Bruers, S. A thermodynamic perspective on food webs: Quantifying entropy production within detrital-based ecosystems. *J. Theor. Biol.* **2007**, *249*, 124–139. [CrossRef]
39. Kleidon, A. Nonequilibrium thermodynamics and maximum entropy production in the Earth system. *Naturwissenschaften* **2009**, *96*, 653–677. [CrossRef]

40. Niven, R.K. Steady state of a dissipative flow-controlled system and the maximuentropy production principle. *Phys. Rev. E* **2009**, *80*, 021113. [CrossRef]
41. Ross, J.; Corlan, A.D.; Muller, S.C. Proposed Principle of Maximum Local Entropy Production. *J. Phys. Chem. B* **2012**, *116*, 7858–7865. [CrossRef]

Concept Paper

Mirror Symmetry Breaking in Liquids and Their Impact on the Development of Homochirality in Abiogenesis: Emerging Proto-RNA as Source of Biochirality?

Carsten Tschierske * and Christian Dressel

Martin Luther University Halle-Wittenberg, Institute of Chemistry, Kurt-Mothes Str. 2, 06120 Halle, Germany; dresselchristian@gmx.de

* Correspondence: carsten.tschierske@chemie.uni-halle.de

Received: 30 May 2020; Accepted: 15 June 2020; Published: 2 July 2020

Abstract: Recent progress in mirror symmetry breaking and chirality amplification in isotropic liquids and liquid crystalline cubic phases of achiral molecule is reviewed and discussed with respect to its implications for the hypothesis of emergence of biological chirality. It is shown that mirror symmetry breaking takes place in fluid systems where homochiral interactions are preferred over heterochiral and a dynamic network structure leads to chirality synchronization if the enantiomerization barrier is sufficiently low, i.e., that racemization drives the development of uniform chirality. Local mirror symmetry breaking leads to conglomerate formation. Total mirror symmetry breaking requires either a proper phase transitions kinetics or minor chiral fields, leading to stochastic and deterministic homochirality, respectively, associated with an extreme chirality amplification power close to the bifurcation point. These mirror symmetry broken liquids are thermodynamically stable states and considered as possible systems in which uniform biochirality could have emerged. A model is hypothesized, which assumes the emergence of uniform chirality by chirality synchronization in dynamic "helical network fluids" followed by polymerization, fixing the chirality and leading to proto-RNA formation in a single process.

Keywords: mirror symmetry breaking; biological chirality; liquid crystals; proto-RNA; networks; compartmentalization; chiral liquids; cubic phases; prebiotic chemistry; chirality amplification; helical self-assembly

---

1. Introduction-Homochirality and Life

Ever since Pasteur revealed the molecular asymmetry of organic compounds in 1848 [1], the origin of the homochirality of biologically relevant molecules has attracted considerable attention. The chirality of organic compounds is due to the tetrahedral structure of carbon with a valence of four bondings to adjacent atoms. Only if all four substituents are different, is the compound chiral and existing as two stereoisomers representing mirror images of each other, differing in their configuration, being either D or L. It is well known that in all existing organisms the carbohydrates exist in the D-form and the amino acids in the L-form (see Figure 1; there are also L-sugars and D-amino acids in biological systems, but if they are involved, they fulfill specific functions, see [2]). This homochirality is considered as a signature of life. In the 4.5 billion years (Ga) of the history of Earth, by 4.2–4.3 Ga Earth has cooled sufficiently to be covered by liquid water. Already by 3.95 Ga, the first signatures of life appear as carbon isotope signatures. In this astonishing narrow window of only about 200–300 million years (0.2–0.3 Ga), the first cells came to existence, meaning that mirror symmetry must have been broken and the genetic code developed during this, on a geological time scale, amazingly short period of abiogenesis [3]. The origin of the homochirality of the molecules of life and the mechanism of the chirogenesis remain an unsolved problem in the hypotheses of development of life [4–7]. The focus of chirogenesis was mostly on the carbohydrates and amino acids being intrinsically chiral, as described in numerous contributions and reviews [8–23]; amphiphiles were also considered [24]. However, there are also biologically relevant molecules, or at least building blocks of such molecules, which are achiral, because they are built up of carbons with only three substituents, and these compounds with trigonal carbons are flat. Among them, the pyrimidine und purine bases (*N*-heterocycles) involved in the nucleic acids form the basis of the genetic code of RNA/DNA, which developed during abiogenesis. Helical self-assembly (for example, predominately single helices for RNA and double helices for DNA) is an important consequence of molecular chirality and simultaneously provides a source of homochirality. It does not require molecular chirality, and therefore it can also be formed under achiral conditions with achiral molecules [25–30]. Helical superstructures are not only formed as crystalline assemblies, fibers and gels [31–38], they are also common in soft matter systems and fluids, where they can occur spontaneously [25,26,39–42] or induced by an internal or external source of chirality [43–45]. Soft self-assembly in fluids is of special interest for chirogenesis, as life most likely developed in aqueous fluids.

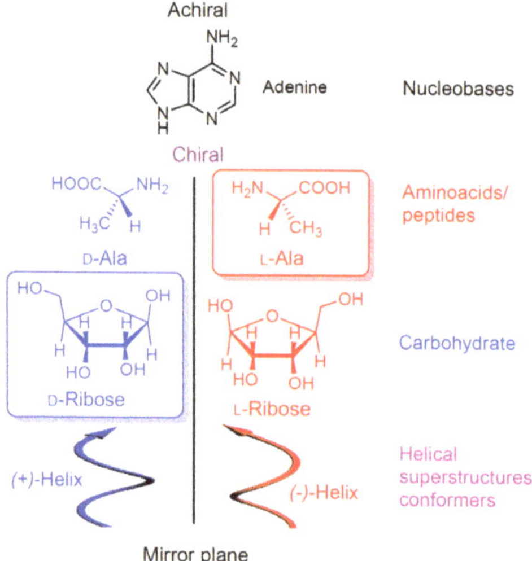

Figure 1. Biologically relevant chiral and achiral molecules and structures.

Herein, the recently discovered spontaneous mirror symmetry breaking in isotropic liquids [25] and in liquid crystalline (LC) cubic phases [26] is discussed under the aspect of their relevance for the emergence of biochirality during abiogenesis. It is postulated that biochirality could have arisen by the spontaneous helical self-assembly of achiral molecules in the liquid state, and that synchronization of the helicity of self-assembled achiral heteroaromatic amphiphiles by the formation of dynamic networks could have been the source of the developing biochirality. It is hypothesized that early forms of RNA could not only have acted as the first information carrier and catalyst in abiogenesis, but could also provided the biochirality.

2. Emergence of Homochirality in Non-Biological Systems-Artificial Chirogenesis

2.1. Racemates vs. Conglomerates

Based on the fundamentals of stereochemistry, under achiral conditions, chiral molecules are always formed from achiral (prochiral) substrates as racemic mixtures of the two enantiomeric forms in

1:1 ratio [46]. In these racemic forms, there are intermolecular interactions between the like enantiomers (D/D and L/L) as well as between unlike pairs (D/L) with different energies. Two different situations can be distinguished depending on the strength of the attractive intermolecular interactions between the like (D/D and L/L) and unlike (D/L) pairs.

As shown in Figure 2a,b, the attractive interactions can be either stronger between the opposite enantiomers, leading to a thermodynamically stable racemate (R-type, Figure 2a), or, if the attractive interactions between like enantiomers are stronger, it leads to conglomerates of the two enantiomorphic pairs D/D and L/L as energy minimum states (C-type, Figure 2b). The first case is the predominating in the crystalline state, observed for 90–95% of the known crystalline compounds [47]. One of the reasons for this is that the entropy gain of mixing contributes to the free energy gain of the racemate, whereas in the case of conglomerate formation, the entropic penalty of demixing must be overcome by the enthalpy gain of the homochiral interactions.

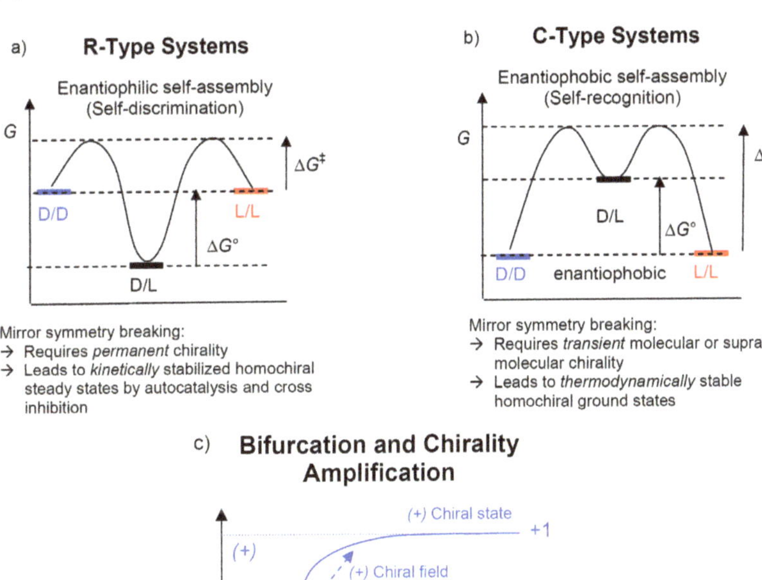

Figure 2. The two systems of relevance for mirror symmetry breaking. (**a**) R-type systems (R = racemate) with stronger attractive forces between the heterochiral enantiomers and low-energy *racemic* ground state (enantiophilic or social self-assembly), mirror symmetry breaking would require an amplification of the excess enantiomer in scalemic mixture or resulting from statistical fluctuations. (**b**) C-type systems (C = conglomerate) with preferred homochiral interactions (enantiophobic or narcissistic self-assembly), leading to low-energy *homochiral* states; this allows the formation of thermodynamically stable *homochiral* ground states, usually representing conglomerates. A third possibility where DD/LL and DL have identical energies is rare and not shown here. A typical feature of both systems is bifurcation into either one of the enantiomeric states, (**c**) shows the general bifurcation scheme; the achiral (or racemic) state becomes instable and chiral statistical fluctuations drive the system to one of the degenerate enantiomeric states; weak chiral polarization close to the bifurcation point transforms the stochastic bifurcation into a deterministic one.

2.2. *R-Type Systems: Mirror Symmetry Breaking by Autocatalysis And Enantiomeric Cross-Inhibition in Chemical Reaction Cycles-Soai Reaction*

In the case of stronger attractive interactions between unlike enantiomers, the racemate is thermodynamically more stable than the homodimers D/D or L/L, see Figure 2a. If a scalemic mixture (i.e., a mixture of the enantiomers being different from 50:50%, the racemate, and from 100:0%, the pure enantiomer) crystallizes, then the racemate is formed first and the excess enantiomer remains in the melt or solution. In this case, uniform chirality can only arise if the ratio of the enantiomers is not precisely 50:50%. However, this can even be the case for the racemates, due to the inherent statistical fluctuations. This means that there is always a tiny statistic deviation from the exact 50:50 ratio obtained in chemical reactions under achiral conditions, either with a minor excess of the D- or the L-enantiomer. The enantiomeric excess (ee) is very small and can be calculated according to ee = $[D] - [L]/[D] + [L] = 0.6743\ (N)^{-0.5}$, where N is the total number of molecules [9,48–54]. This provides the basis of the mechanism of spontaneous emergence of homochirality in chemical systems proposed in 1953 by F. C. Frank [55]. It is based on autocatalysis and inhibition by the formation of an inactive meso-form by D + L pairing (enantiomeric cross-inhibition), leaving the pure excess enantiomer acting as the autocatalyst for its own reproduction. The statistical fluctuations lead, in a stochastic way, either to one or the opposite chiral form with equal probability (bifurcation, see Figure 2c). An imbalance arises by chiral dopands or chiral forces (both summarized here as chiral fields) such as circular polarized light and chiral surfaces, which bias one sense of chirality (dotted arrows in Figure 2c) [56]. Soai et al. [57,58] found the first highly efficient asymmetric autocatalysis (non-linear chirality transfer [59]) with a strong positive non-linearity of chirality transfer (amplification of chirality), by using the reaction between pyrimidine-5-carbaldehydes with diisopropyl zinc (Figure 3) [57]. A huge number of weak sources of chirality have been tested, including cryptochiral compounds [60], isotopomers [61] enantiotopic surfaces of crystals [62] and statistical fluctuations [63,64]. An enantiomeric excess (ee) as low as 5×10^{-5}% was found to be sufficient to ensure almost complete enantioselection of > 99.5% in only three autocatalytic cycles (Figure 3). Though this specific reaction is very unlikely to have played any role in the development of homochirality in nature, it is an important proof of principle. This process can produce chiral molecules in an isotropic solution by chemical reactions under achiral conditions (stochastic) or under the influence of a weak chiral field (deterministic). Usually, in the experiments, the chirality of the products is fixed by crystallization, which makes them long-term stable. However, if kept in solution racemization is likely to occur by the formation of the thermodynamically more stable racemic state and the thus obtained chirality is instable on a geological time-scale [65]. This means that the chiral fluids of R-type systems represent metastable states which must be maintained by continuous autocatalytic cycles and a continuous input of energy, as described in a number of papers [16,17,55,66,67].

2.3. *C-Type Systems: Spontaneous Homochirality*

2.3.1. Spontaneous Homochirality by Phase Transitions-Viedma Ripening

Cross-inhibition and autocatalysis are not required for achieving homochirality if the homochiral attractive interactions become preferential over the heterochiral (C-type systems, see Figure 2b) [68–72]. In this case, the homochiral state represents the spontaneously formed energy minimum state and homochirality can emerge spontaneously, leading to conglomerates during crystallization of racemic mixtures, as already shown by Pasteur's famous work [1]. The typical outcome is a conglomerate of both coexisting enantiomeric forms, i.e., only a locally symmetry broken state is formed, but recently methods have been developed to shift this racemic mixture to only one of the enantiomeric states, thus leading to complete mirror symmetry breaking.

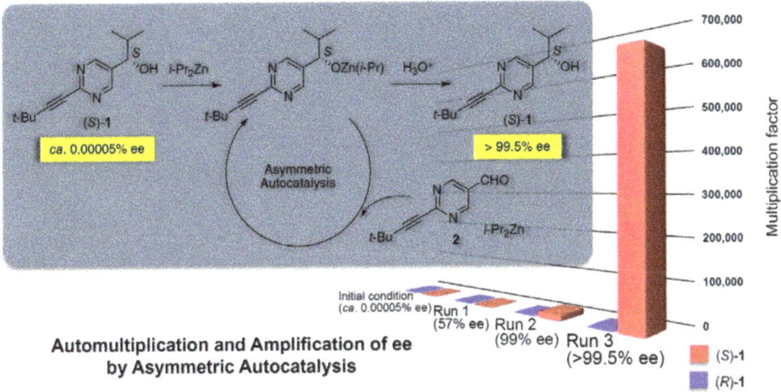

Figure 3. Asymmetric autocatalytic chirality amplification in the Soai reaction leads in only three cycles from (*S*)-**1** with an ee of only 5×10^{-5}% ee to an almost enantiomerically pure compound (*S*)-**1** with > 95% ee, representing an extreme case of deterministic mirror symmetry breaking in an R-type system [57]. The obtained homochiral enantiomeric product is stabilized and racemization is kinetically inhibited by storage of the obtained compound in the crystalline state. Reprinted from [57], Copyright (2014), with permission from Wiley.

The most prominent example of such a process is the Viedma ripening, which allows the complete deracemization of conglomerates by grinding or thermal cycling with a coexisting solution (Figure 4b) [73–75]. Without going into details, this process is based on the dissolution of smaller crystals and the growth of the lager, combined by their fast racemization in the coexisting solution. In this case, enantiomerization of the minor enantiomer is essential for achieving uniform chirality. This racemization must be sufficiently fast and therefore in most cases requires catalysis for permanently chiral molecules [76]. This can also be used in chemical reactions, where a fast equilibrium between the enantiomeric products and the achiral educts leads to the crystallization of the product in 100% ee, representing a new method of absolute asymmetric synthesis [77–80]. Likewise, achiral molecules (e.g., $NaClO_3$ [73,74]) and transiently chiral molecules with chiral low-energy conformations [42] (e.g., benzil) can crystallize with the formation of only one enantiomeric form of chiral crystals (Figure 4a,c) [81]. A fast racemization kinetics is the basis of the spontaneous mirror symmetry breaking in the case of the C-type systems (Figure 2b), whereas for the R-type systems (Figure 2a) racemization leads to the loss of chirality. Moreover, in all these cases, the mirror symmetry broken state once developed is trapped by crystallization in a 3D lattice which provides a denser packing and freezes the dynamics. This, together with the high degree of cooperativity of the non-covalent interactions between the molecules in the crystalline 3D lattice enhances the enantiomerization energy barrier. Therefore, even for fast racemizing chiral molecular conformers, the chirality sense becomes synchronized and fixed in the crystalline state (Figure 4c). However, in dilute solutions, the interactions with the achiral solvent molecules become the dominant intermolecular interaction (solvatation) and the attractive interactions between like enantiomers cannot stabilize the system. Therefore, in solution, the unfavorable entropy of the homochiral system leads to racemization in timescales depending on the activation energy of racemization. The situation is different for polymers, where the chiral units are covalently fixed to one other and stabilize each other, and therefore the favorable like interactions are also retained in solution, as long as the polymer is not hydrolyzed [82].

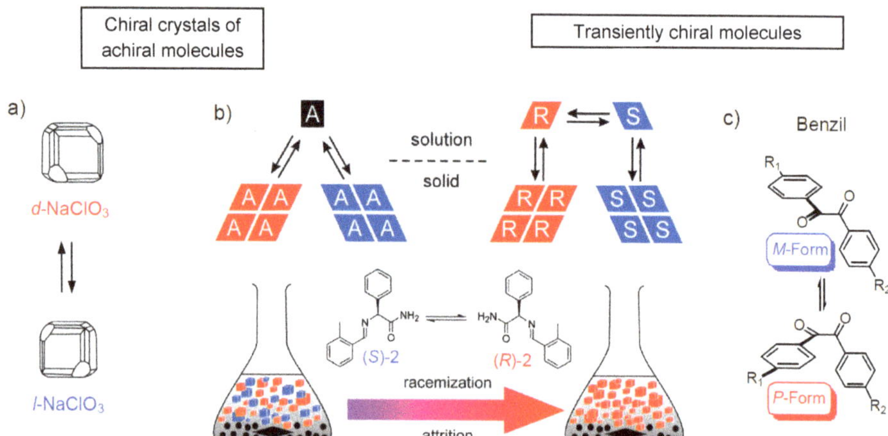

Figure 4. Viedma deracemization. Panel (**b**) shows the principle scheme for the transformation of a racemic conglomerate of crystals of an achiral (A) or racemic chiral compounds (R ↔ S) in solution by mechanical grinding. [77] (**a**) shows the example of chiral crystals formed by achiral molecules and (**c**) an example of a transiently chiral molecule, separated by this process into enantiomorphic crystals; in the crystals, the conformers are chirality synchronized. Note that the color scheme indicating the chirality sense is in this figure opposite to that used in the other figures. (**b**) is reprinted from [77], Copyright (2009), with permission from Wiley-VCH.

2.3.2. Spontaneous Mirror Symmetry Breaking in Liquids

The above-mentioned stabilizing effects of the crystalline state are absent in the melted isotropic liquid state. However, for C-type liquids, the spontaneous bifurcation into two liquids with opposite optical activity was identified by computer simulations [65,68]. It was shown that it takes place if *(i)* the enthalpic gain of deracemization exceeds the entropic penalty of demixing of the enantiomers by $\Delta H = 2k_BT$ (Figure 5), and *(ii)* if there is a fast enantiomerization kinetics. Both together lead to a synchronization of the chirality sense. The resulting enantiomeric liquids are scalemic and increase in enantiomeric purity with further rising ΔH.

As the stronger attractive interaction between the like enantiomers is likely to lead to a closer packing, the conglomerate formation takes place by a decrease in volume, and therefore it was recently proposed that spontaneous resolution in such liquids could be triggered under sufficiently high pressure [83,84]. Nevertheless, attempts to achieve spontaneous separation of enantiomers in the liquid state have failed [85,86] until, in 2014, we discovered the first experimental observation of spontaneous mirror symmetry breaking in the liquid state of transiently chiral molecules (compound **3** in Figure 6), even without any applied pressure [25]. In the meantime, this phenomenon was observed in several other classes of compounds, shown in Figure 6 [40,42,87–90]. In the following sections, the possible reasons for this observation and the potential impact of this finding for the emergence of uniform biochirality will be discussed.

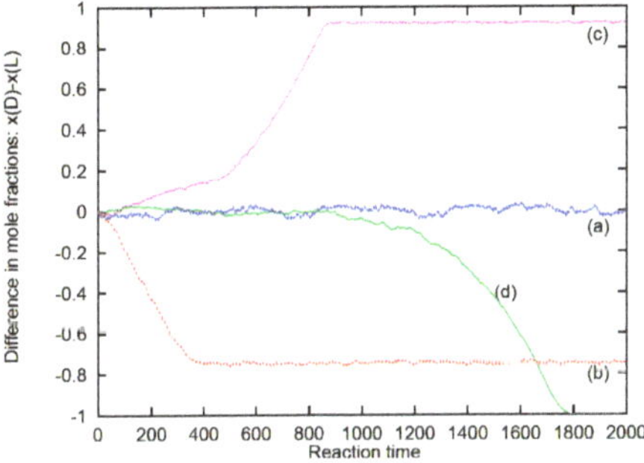

Figure 5. Enantiomeric excess as a function of time in a molecular dynamic system with isomerisation kinetics and for different enthalpy gains for the chirality synchronization process with (**a**) $|\Delta H| = RT$, (**b**) $2RT$, (**c**) $3RT$ and (**d**) ∞; reprinted from [65].

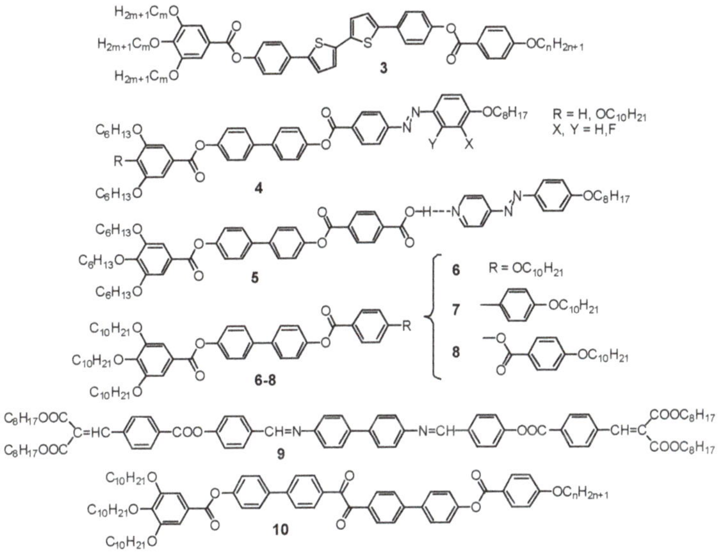

Figure 6. Molecular structures of compounds capable of mirror symmetry breaking in the isotropic liquid state (Iso$^{[*]}$ phases) [25,26,87–90].

3. From Crystals via Liquid Crystals to Liquids

Liquid crystals (LCs) represent intermediate states between ordinary liquids and the crystalline state combining order and mobility; the order can be either orientational or positional or a combination of both [91–93]. This leads to a huge variety of LC states, a small selection of the most important phase structures, those of relevance for the discussion herein, are collated in Figure 7a–f. The simplest LC phase is the nematic phase (Figure 7a), which has only orientational order and makes the physical properties of the liquid direction-dependent [94]. Emerging positional order can occur in one, two or all three spatial dimensions, leading to lamellar (smectic), columnar and bicontinuous cubic (Cub_{bi}) phases, respectively (Figure 7b–f). The most complex LC phases are the Cub_{bi} phases, which develop at the transition between lamellar (1D) and columnar (2D) self-assembly [95–102]. These Cub_{bi} phases with a 3D lattice represent networks of branched columns. Due to the intrinsic combination of mobility and order, all three modes of LC self-assembly are of relevance for biosystems [103–109]. The lamellar structures are the basis of all cell membranes [110,111]: columnar phases allow the dense packing of DNA in the nucleus [112–115] and cubic phases are involved in cell fusion and cell fission processes [95,102]. The positional order is mainly a consequence of the segregation of incompatible molecular building blocks into separate nano-spaces (microsegregation or nanosegregation [93,97,116]), i.e., these compounds represent amphiphiles, which are composed of two chemically connected incompatible parts (Figure 7g) [97]. However, the amphiphilicity is not restricted to the well-known polar/lipophilic amphiphilicity, as commonly known for lipids and their aqueous systems (Figure 7h); it can arise from any incompatibility of intermolecular interactions, and the second most common is the incompatibility between rigid polyaromatic and flexible aliphatic building blocks [93,97,117]. In all cases, this segregation creates interfaces which determine the morphology of the aggregates and the dimensionality of long-range positional order on an nm-scale (Figure 7b–f). The remarkable feature of all LCs, which distinguishes them from the solid-state crystals, is the absence of any fixed position of the individual molecules, which is identified by the typical diffuse scattering around 4.5 nm observed in their X-ray diffraction patterns. This means that despite the presence of long-range order, the individual molecules are still mobile and thus can be considered as ordered fluids. However, due to the order, the molecules are not completely independent, but more or less fixed in lamellar or columnar aggregates or even in networks (Cub_{bi} phases), and this provides a confinement for the molecules and cooperativity of the intermolecular interactions in one, two or even three directions [42,117,118]. Consequently, mirror symmetry breaking is supported and stabilized in the LC state compared to the isotropic liquid state, and there are numerous recent reports of spontaneous mirror symmetry breaking in LCs. These have been summarized and discussed in previous reviews [40–42,119].

We focus here only on those observed in the Cub_{bi} phases, which are of relevance for the following discussions [95,96]. Bicontinuous means that there is a continuum and additional continuous networks, independent of the actual number of these networks. The networks are distinct by the valence of their junctions (mostly three, but also four and six [96]) and the number of networks per unit cell (in most cases two [96,120], but in some cases also one [121,122] or three [123]). Figure 7d–f shows the case of networks with three-way junctions which are the most common, namely the single gyroid, the double gyroid and the more complex *I*23 phase. To date, only the double network (the gyroid, *Ia*3*d*) and the triple network (*I*23) [123] are known for liquid crystalline Cub_{bi} phases with three way junctions (Figure 7d,f). The continuum is, in most cases, either formed by a solvent (in most cases water) in the so-called *lyotropic* systems, or by flexible chains, in most cases alkyl chains, in the solvent-free *thermotropic* systems. The fluidity of solvents or chains contributes significantly to the mobility and to the segregation of networks and continuum. However, in the networks the organization is liquid-like, not crystal-like, i.e., liquid crystalline Cub_{bi} phases are dynamic and therefore very different from the networks in gels which are crystallized [29,31,32].

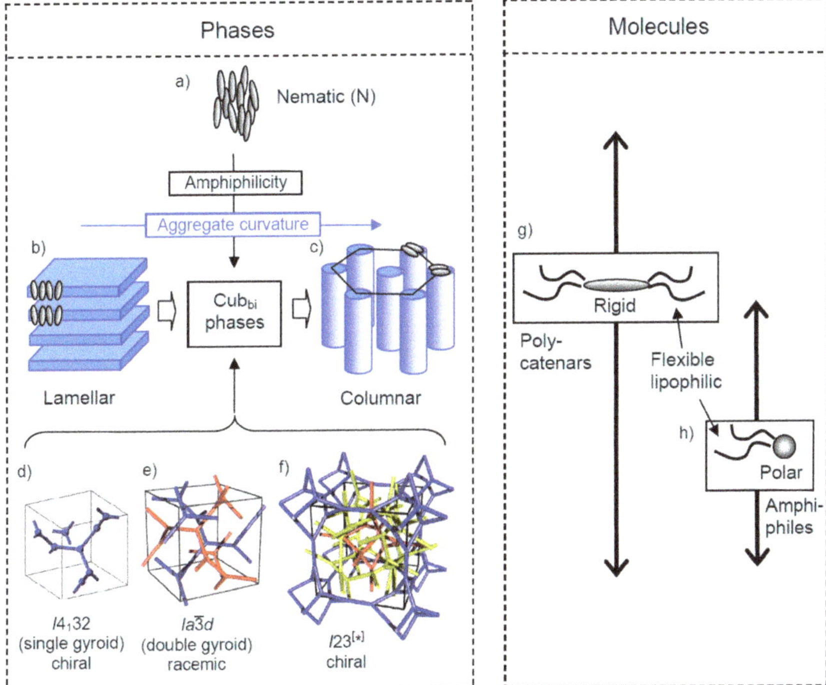

Figure 7. Development of liquid crystalline (LC) phases depending on molecular amphilicity and supramolecular aggregate curvature. (**a**) Nematic phase formed by non-amphiphilic rod-like molecules showing exclusively orientational order while retaining liquid-like fluidity. (**b**–**c**) Transition from lamellar via bicontinuous cubic (Cub_{bi}) phases to columnar phases by increasing interface curvature between the nano-segregated compartments, as observed for polycatenars compounds. The Cub_{bi} phases are specified in d-f showing cubic phases with three-way junction networks. (**d**) The single gyroid is yet unknown for LC systems; (**e**) the double gyroid is the commonly observed Cub_{bi} phase for a wide range of self-assembled systems, ranging from lyotropic systems to bulk self-assembly of low molecular weight and polymeric amphiphiles; (**f**) the complex triple network structure with *I*23 lattice is exclusively formed by rigid π-conjugated molecules crystals, replacing the *Ia*3*d* phase for a distinct range of chain volume. For the single- and double-network structures, four- and six-way junctions are also possible, which are not shown. Panels (**g**,**h**) show two typical molecular structures forming Cub_{bi} phases, the arrows indicate the LC phase types covered by these compounds. (**f**) is reprinted from [123], Copyright (2020), with permission from RSC.

These Cub_{bi} phases are special among the LC phases for several reasons. Firstly, due to the cubic symmetry, they are isotropic like ordinary liquids. Due the optical isotropy, they do not show any linear birefringence, and hence, optical activity and mirror symmetry breaking can be easily be detected by the rotation of the plane of linear polarized light. Under a polarizing microscope, which can be considered as a polarimeter with spatial resolution, a chiral conglomerate appears as a pattern of dark and bright domains for the areas with opposite chirality after a slight rotation of one polarizer out of the 90° crossed orientation by 1°–10° in a clockwise or anticlockwise direction. The brightness is inverted by inverting the twist direction of the polarizers (see Figure 8e,f) [26]. Secondly, Cub_{bi} phases represent supramolecular network structures in a continuum, and networks are responsible for the development of complexity and emergence of new properties [124]. In the case discussed here, the self-assembled molecular 3D network structure leads to the emergence of mirror symmetry breaking due to the improved pre-organization of the involved molecules and the thus increasing intermolecular interactions and arising cooperativity [116,118]. The network structure leads to a transmission of chirality in all three spatial directions. Moreover, the inherent mobility allows a fast enantiomerization kinetics, which is required for mirror symmetry breaking in fluids.

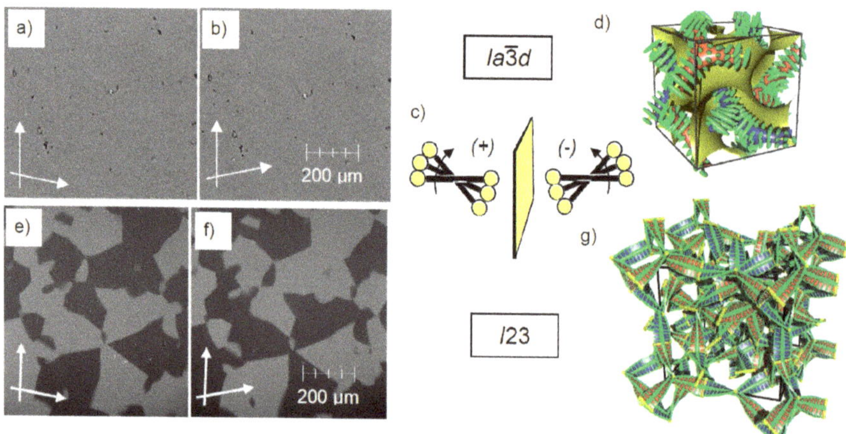

Figure 8. Mirror symmetry breaking in the Cub_{bi} phases. In (**a**,**b**) and (**e**,**f**) the textures as observed under a polarizing microscope with the indicated orientation of the polarizers (white arrows) are shown [26], and (**c**,**d**,**g**) show the models of molecular organization. (**c**) shows the helical twist of the molecules along the networks, which can be right- or left-handed. (**d**,**g**) show one unit cell of the cubic phases; the yellow interfaces in (**d**) represent the minimal surfaces separating the two networks; the green rods in (**d**) and in (**g**) the yellow rods and the thin green lines parallel to these rods indicate the rod-like units organized in a helical manner [123]. Panels (**a**,**b**,**d**) show the achiral double network Cub_{bi} phase with *Ia3d* space group. Panels (**e**–**g**) show the chiral triple network Cub_{bi} phase with *I*23 space group; adapted from [26], (**g**) is reprinted from [123], Copyright (2020), with permission from RSC.

4. Mirror Symmetry Breaking in Cubic Network Phases

4.1. Racemic Ia3d Phases and Chiral Conglomerate Type I23 Phases

Though there are numerous different materials capable of forming thermotropic or lyotropic Cub_{bi} phases [92,93,99,100], the focus is here on the thermotropic Cub_{bi} phases formed by compounds having flexible chains at one or both ends of a rod-like polyaromatic core (see Figure 6). In the Cub_{bi} phases of these so-called polycatenar molecules [125–127], the rods are organized in networks and the terminal chains form the continuum between them [26,42]. Within the networks, the rods tend to be organized, on average, parallel to each other and perpendicular to the networks (Figure 8d,g). However, due to the larger space required by the end-chains compared to the rods, they cannot align perfectly parallel, but assume a helical twist (Figure 8c), which is transmitted by the junctions throughout the networks. In this way, dynamic supramolecular helix networks develop, with the helix sense being synchronized along the networks (Figure 8d,g) [26,42].

Depending on the number of networks, the symmetry of the cubic lattice is different. The *Ia3d* lattice is formed by two enantiomorphic helical networks with three-way junctions being twisted to each other by 70.5° (Figures 7e and 8d) [26]. This means that the networks themselves are homochiral and two networks with opposite handedness are combined in the *Ia3d* lattice, thus becoming overall achiral, despite the presence of a helical superstructure in the networks. The second type of cubic phases observed for this kind of compounds has the space group *Im3m*, which is achiral and formed by three networks (Figure 8g) [26,123]. In this case, the network chirality cannot compensate and there is a synchronization of the helicity in all three network, leading to the chiral space group *I*23.

All Cub_{bi}/*I*23 type phases investigated to date show spontaneous mirror symmetry breaking under all conditions, though they are formed by achiral molecules [26,40,42,87–90,128–133]. The chirality is detected as described above by investigation under the polarizing microscope by the formation of a conglomerate composed of dark and bright domains due to the rotation of the polarization plane of the light into opposite directions (Figure 8e,f) [26,40,42]. The high optical rotation is about 1–2°/10 μm (corresponding to about 10,000–20,000°/dm), mainly due to the exciton coupling between the twisted π-conjugated rods [134,135]. In contrast, the achiral *Ia*3*d* phase remains uniformly dark after rotating one of the polarizers in either direction, confirming that it is optically inactive (Figure 8a,b). This shows that the network formation in the LC Cub_{bi} phases, transmitting chirality and disfavouring helix reversals [136], is a powerful tool for achieving mirror symmetry breaking under thermodynamic equilibrium conditions and for the long-term stabilization of the chirality synchronized helices in the LC network phases with cubic [26,88,89,128–133] or non-cubic symmetry [137,138]. It is noted that the first report on mirror symmetry breaking in cubic phases came from Kishikawa et al. though the structure of the cubic phase and the origin of chirality remained unclear at that time [139]. Overall, the *Ia*3*d* phase can be considered as an R-type and the *I*23 phase as a C-type Cub_{bi} system. Only in C-type systems can mirror symmetry breaking be expected to be achieved by transiently chiral molecules or aggregates in fluid systems under thermodynamic control (Figure 2a,b). Such C-type *I*23 triple network structures have, to date, only been observed for systems involving sufficiently long rigid polyaromatic units [26], whereas flexible amphiphiles of any kind prefer the organization in the R-type double gyroid (*Ia*3*d*) cubic phases [93,96,99–101].

4.2. Conglomerate Formation vs. Complete Mirror Symmetry Breaking in Cubic Pases

As shown in Figure 2c mirror symmetry breaking leads to a bifurcation with stochastic outcome, i.e., either one or the other enantiomer becomes dominating, depending on the statistical fluctuations in the considered system. For fast isomerizing systems, these fluctuations modulate locally in space and with time [48,52,56]. Therefore, as shown in Figure 8e,f, conglomerates of chiral domains with opposite handedness are usually observed, not a uniform chirality of the complete sample. Thus, mirror symmetry breaking is usually only local and takes place under compartmentalization of the sample, where the size of the chiral domains depends on many parameters. Strictly speaking, this is only a local mirror symmetry breaking which, on a larger macroscopic scale, still retains a non-symmetry broken overall racemic state.

Symmetry breaking becomes complete, i.e., only one sense of chirality can be observed in the considered system, if the seed formation at the Iso-*I*23 transition is slow and the following growth of the uniform domains is sufficiently fast that only one seed determines the chirality sense of the complete sample. However, whether mirror symmetry breaking is considered as complete or only local depends on the size of the actually investigated system. In the case of the Cub_{bi} phases, the investigations are usually carried out with 10–50-μm-thick films between two glass substrates and, in this case, the size of the uniformly chiral areas can easily become as large as one to few square millimeters. However, on a larger length scale, conglomerate formation is still observed as soon as more than only one initial seed is formed, and its chirality sense is determined by the local fluctuations (stochastic symmetry breaking). However, the development of uniform chirality is easily triggered by tiny internal or external chiral fields (Figure 2c) [25]. Tiny sources of chirality by (often not detectable) traces of chiral impurities or chiral physical forces can thus lead to only one or the other enantiomer in the whole investigated system, leading to complete deterministic mirror symmetry breaking.

The parity violation is a potential universally acting chiral field, but its effect on mirror symmetry breaking has not yet been unambiguously confirmed experimentally [140] and is considered as unlikely [141]. This effect is much smaller than k_BT and therefore has no measurable influence, at least in non-polymeric systems. Overall, chiral conglomerate formation is considered as an indication of the capability of the considered system of spontaneous mirror symmetry breaking in a finite system.

5. Experimental Demonstration of Mirror Symmetry Breaking in Isotropic Liquids

5.1. Local Mirror Symmetry Breaking by Conglomerate Formation

The first experimental proof of a spontaneous mirror symmetry breaking in the liquid state, occurring even in the absence of any applied pressure and at temperatures compatible with the conditions of abiogenesis, was observed for liquid phases occurring in the vicinity of the Cub_{bi} phases of polycatenar molecules, like compounds **3** [25]. The molecules forming these mirror symmetry broken liquids are achiral (Figure 6), but can assume energy minimum chiral conformations which are in a rapid equilibrium (transiently chiral molecules, Figure 9a,b) [42].

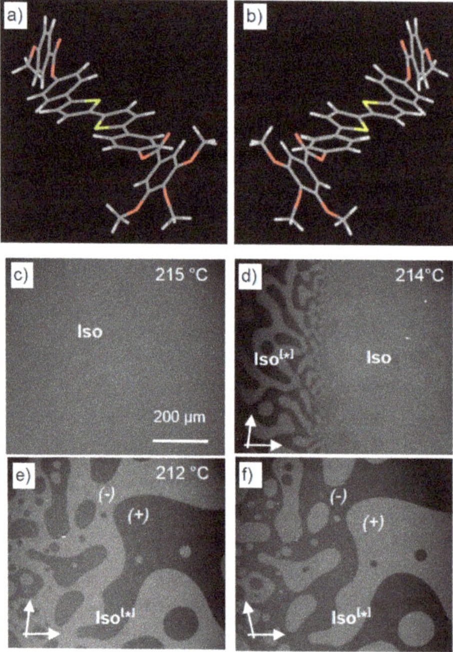

Figure 9. The liquid phases of the polycatenar compounds **3**. (**a**,**b**) Molecular models showing the helical conformers of a model compound **3** with $m, n = 1$; (**c**–**f**) Transition from the achiral isotropic liquid to the mirror symmetry broken $Iso^{[*]}$ phase as observed for compound **3** with $m, n = 6$ on cooling between slightly uncrossed polarizers; (**c**) achiral Iso phase, (**d**) transition Iso-$Iso^{[*]}$ at 214 °C showing spinodal decomposition at the phase boundary (there is a temperature gradient in the sample with slightly higher temperature at the right), and (**e**,**f**) fully segregated chiral domains in the $Iso^{[*]}$ phase at 212 °C, inverting the brightness by uncrossing the polarizers either by clockwise or anticlockwise rotation; adapted from [25].

As shown in Figure 9c–f, the optically isotropic liquid phase (Iso) spontaneously becomes mirror symmetry broken at a phase transition by the formation of a conglomerate of liquid domains with opposite signs of optical rotation ($Iso^{[*]}$ phase). This process is completely reversible without any visible supercooling effect, even at high heating/cooling rates. The spinodal decomposition confirms the first-order nature of the phase transition. At this phase transition, there is no visible change in the viscosity, i.e., the $Iso^{[*]}$ phase is also a true liquid which flows under gravity, as shown in a video attached as supporting information to [25].

The liquid state is additionally confirmed by X-ray scattering (XRD). The XRD patterns show only a diffuse scattering in the small- as well as in the wide-angle range, excluding the formation of a cubic phase [25]. The shape of the diffuse small-angle scattering changes continuously and becomes continuously narrower across the Iso-Iso[*] transitions, but without any visible discontinuity [25]. This supports the liquid state of the material and indicates a spontaneous separation process, leading to a macroscopic conglomerate of two chemical identical, but immiscible liquids differing only in the direction of the rotation of the plane of polarized light [25]. In this case, spontaneous mirror symmetry breaking takes place under thermodynamic equilibrium conditions, leading to a conglomerate of two chiral liquids which is stable, even at temperatures as high as >200 °C. This demonstrates for the first time that in dynamic systems chirality can develop spontaneously from *achiral* molecules under thermodynamic control [25]. In the meantime, this was found for a number of different polycatenars molecules (Figure 6) [25,88,89,130], hydrogen-bonded aggregates **5** and even for simple heterocyclic compounds with relationships with nucleobases (see Section 6 below). Even photoisomerizable compounds were found, for which the chirality can be switched on and off by non-polarized visible light (compounds **4**, **5** in Figure 6) [88–90].

5.2. Total Mirror Symmetry Breaking in Isotropic Liquids

Typically, the formation of the chiral Iso[*] phase takes place by a spinodal demixing at the Iso-Iso[*] transition where, due to local fluctuations, small left- and right-handed domains spontaneously emerge in 1:1 ratio (Figure 10a). The domains rapidly grow and fuse with the formation of larger domains with uniform chirality as a result of the reduction in the interfacial areas between the oppositely chiral domains and the minimization of the curvature of these interfaces (see Figure 10b).

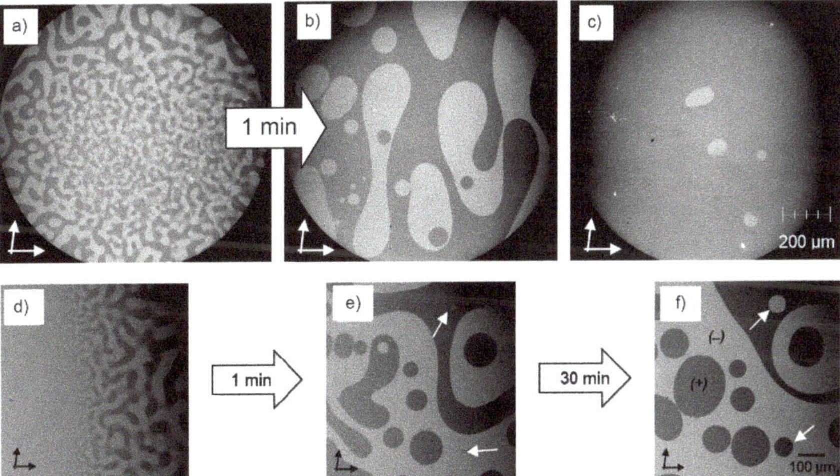

Figure 10. Growth of the size of the chiral domains in the Iso[*] phases of compounds **3** with time. (**a**,**b**) After the transition from the achiral *Ia3d* phase to the mirror symmetry broken Iso[*] phase of compound **3** with *m*, *n* = 6 on heating; (**a**) developing chiral domains immediately after the phase transition *Ia3d*- Iso[*] at *T* = 205 °C and (**b**) after one min at *T* = 210 °C; (**c**) shows an example for the spontaneous imbalance of left- and right-handed domains associated with Iso[*] phases occurring besides a chiral *I*23 phase (compound **3** with *m* = 10 and *n* = 8); (**d**–**f**) show the Iso-Iso[*] transition on cooling of compound **3** with *m* = *n* = 6, (**d**) at the phase transition at *T* = 212 °C and (**e**,**f**) after growth for one and 30 min at *T* = 190 °C, respectively; note that the growth of the small circular domains which is indicated with arrows (previously unpublished results).

Toxvaerd proposed that these chiral liquids show an intrinsic tendency to destabilize the racemic state and only one homochiral state will appear when one of the chiral domains encapsulates the other [19,68]. The argument is that the enantiomerization takes place at the interface, whereas the chirality synchronization inside the chiral domains slows down the kinetics, which enhances the segregation of the enantiomers, i.e., the molecules at the surface are activated. In a droplet of an L-rich domain in the D-rich continuum, the activated molecules at the interfaces have a larger probability to assume the chirality sense of the molecules on the convex side of the droplet (outer side towards the continuum), which experience an excess of D-enantiomer and the droplet will shrink and finally disappear [19]. However, this could not be confirmed experimentally. In our case, the opposite was found: the droplets grow with time (Figure 10a,b,d–f) and then fuse to larger domains. This takes place in the D-rich and L-rich regions, and the growth process slows down with growing domain size and decreasing curvature of the inter-domain interfaces (Figure 10e,f). After a certain domain size is reached, no further changes can be observed. This means that the system apparently tends to retain the overall racemic conglomerate state (at least in laboratory timescales) as long as there is no imbalance right from the beginning due to any kind of a biasing chiral field. However, this cannot be stated with certainty, due to the dramatic slowdown of the agglomeration process as the interfacial curvature between the domains decreases. Hochberg and Zorzano have simulated the effect of spatio-temporal fluctuations on the Frank chiral amplification model and obtained two solutions, one leading to uniform chirality and absolute symmetry breaking (Figure 11a) and a second one leading to a racemic state with a linear phase boundary (Figure 11b) [52]. The latter was found to occur in two dimensional systems. Therefore, it must be considered that the experimentally investigated system is a relatively thin film (10–50 μm) between two substrates and, in this case, it could behave like a two dimensional system which stacks in the racemic case, whereas in true 3D systems, homochirality might be the final outcome. It is noted here that, once a homochiral liquid has been formed (either under a chiral field or spontaneously after transition from a homochiral Cub_{bi} phase, see below), this liquid is stable and no racemization could be observed.

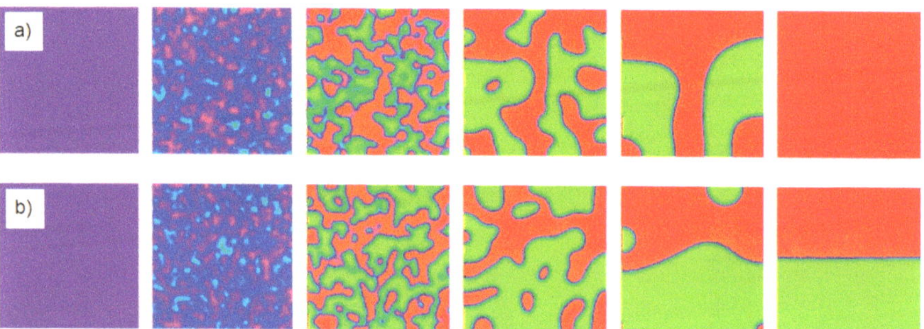

Figure 11. Results of simulations using the exact Langevin equations. Starting from a homogeneous racemic condition (left) the local fluctuations drive the system (**a**) to a homochiral state (right) or (**b**) to a state with spatial segregation (right); time runs from left to right. Reprinted from [52], Copyright (2006), with permission from Elsevier.

5.3. *Effects of Dilution on Mirror Symmetry Breaking in Isotropic Liquids*

It is also worth noting that these liquid conglomerates tolerate the addition of achiral molecules, such as, for example, *n*-alkanes, which can even expand the Iso[*] phase temperature range [25]. As shown in Figure 12, up to 50 mol% of a long-chain *n*-alkane ($C_{32}H_{66}$) is tolerated. This is important with respect to the discussion of the possible impact of chirality synchronization in the isotropic liquid state on the emergence of homogeneous chirality in aqueous prebiotic fluids.

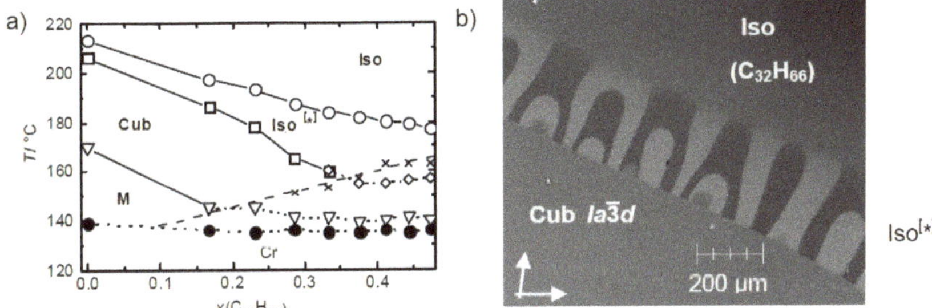

Figure 12. Effect of an achiral hydrocarbon solvent on the Iso[*] phase range of compound **3** ($m = n = 6$, see Figure 6); (**a**) shows the development of the phase ranges depending on concentration of the hydrocarbon $C_{32}H_{66}$ and temperature and (**b**) shows the contact region as observed under a polarizing microscope at $T = 200$ °C (orientation of polarizer and analyzer indicated by arrows), showing the development of the mirror symmetry broken Iso[*] phase between the achiral *Ia*3*d* phase (bottom) and the achiral hydrocarbon solvent (top); reprinted from [25].

5.4. Chirality Amplification in Isotropic Liquids

Under strictly achiral conditions, the area ratio of the domains is 1:1 in the liquid conglomerates, but chiral dopands with a concentration as low as only 10^{-8} mol% can shift the ratio of the domains to a clearly visible imbalance, whereas 10^{-3} mol% is sufficient to produce homochirality [25]. Thus, these conglomerates have a huge chirality amplification power, comparable with those observed for the Soai reaction, and weak chiral influences can easily lead to homochirality. The effect of chiral dopands is the most efficient if it is present at the phase transition temperature (Iso-Iso[*] or Iso-*I*23), i.e., close to the bifurcation point (Figure 2c); below this transition, the effects become much smaller.

5.5. Total Mirror Symmetry Breaking in the Cub_{bi}/*I*23 Phase

Complete mirror symmetry breaking (homochirality) in the absence of any chiral field can be achieved by a phase transition from the chiral Iso[*] phase to a chiral cubic *I*23 phase if the formation of the seeds of this cubic phase is a slow process and the transition Iso[*]-Cub_{bi}/*I*23 is significantly faster, as shown in Figure 13a–d. In the ideal case, a single seed, which develops either in a (+)- or (−)-liquid domain, leads to uniform chirality in the whole sample, thus leading to complete mirror symmetry breaking. Although the growth of the *I*23 phase is faster in the regions with the same chirality, it moves the domain boundaries into the areas with opposite chirality and thus leads to an inversion of the chirality in these regions (see Figure 13a→b→c), finally providing uniform chirality (Figure 13d). Although this growth is slower, because it requires helix inversion at the moving interfaces, the whole process shown in Figure 13a–d takes place within less than 10 s. Occasionally homochirality can also be found at the transition from the achiral isotropic liquid to the *I*23 phase on cooling, which is, however, difficult to detect, because in the absence of the conglomerate texture, no significant optical change is visible at the phase transition.

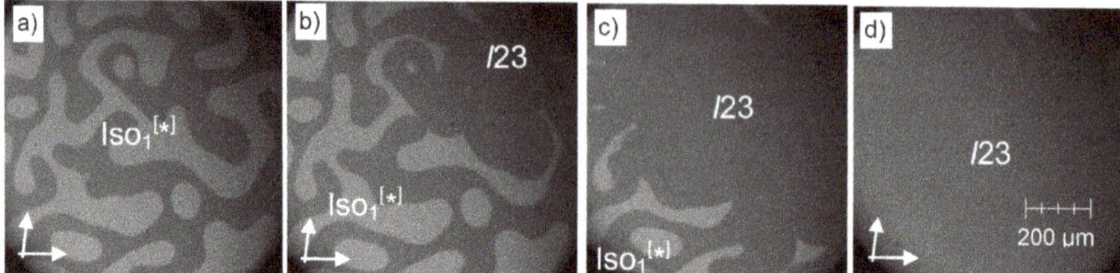

Figure 13. Spontaneous development of uniform chirality at the $Iso^{[*]}$-to-$Cub_{bi}/I23$ transition, shown for compound **3** ($m = n = 10$) and as observed under a polarizing microscope at $T = 175$ °C (orientation of the polarizers is indicated by arrows), depending on time; (**a**) shows the $Iso^{[*]}$ phase at $t = 0$ s; (**b**) formation of the *I*23 phase starts in the dark region at the top right, (**c**) the *I*23 phase area grows towards the bottom left and (**d**) covers the complete area of view after < 10 s; note that the optical rotatory power does not change at this transition, meaning that the local structure in the $Iso^{[*]}$ phase is likely to be the same as in *I*23; see also the video associated as supporting material to [26].

5.6. $Cub_{bi}/I23$ Phase-Assisted Total Mirror Symmetry Breaking in Isotropic Liquids

In some cases, it was observed that the formed $Iso^{[*]}$ phase occurring beside the chiral cubic *I*23 phase is spontaneously shifted to one chirality, even in the absence of any chiral field (Figure 10c). This observation results from two different effects. First, the formation of the chiral cubic *I*23 phase may tend easily towards an imbalance of left- and right-handed domains, in most cases only slightly, but sometimes almost completely, as shown in Figure 13d, which creates a chiral field for the developing $Iso^{[*]}$ phase. The other effect is the storage of the chiral information in the liquid state. It appears that the chiral information once gained in the *I*23 phase is stored either in clusters of networks persisting in the isotropic liquid phase range close to the $Iso^{[*]}$-Iso phase transition temperature, or in thin films at the surfaces. This range of the achiral isotropic liquid phase with a local network structure is designated here as Iso_1. The chiral memory is completely lost within few minutes in the achiral Iso phase at >20 K above the phase transition and in this case upon cooling the racemic conglomerate composed of equal areas of the chiral domains is formed again. As the temperature difference to the $Iso^{[*]}$-Iso_1 phase transition is reduced, the time required for complete loss of chiral memory increases. Even in a certain temperature range of the achiral isotropic liquid phase close to the $Iso^{[*]}$-Iso_1 transition, the chirality is memorized, so that, on cooling back to $Iso^{[*]}$, either a uniformly chiral $Iso^{[*]}$ phase is formed again or an equal or non-equal distribution of left and right domains is found. The outcome depends on the temperature difference to the $Iso^{[*]}$-Iso_1 phase transition temperature and on the time; it is kept at that temperature in the achiral Iso_1-phase. In this way, the interaction of the spontaneous formed chiral cubic phase with a ratio of left- and right-handed domains unequal the racemic 1:1 state, and memorizing this chiral information in the liquid state, generates a chiral field which easily leads to symmetry breaking over larger areas, or maybe total mirror symmetry breaking if the cycle $Cub_{bi}/I23$-Iso_1-$Iso^{[*]}$ is repeated.

5.7. Complete Mirror Symmetry Breaking in Isotropic Liquds Due to Compartmentalization

Another interesting observation made in the $Iso^{[*]}$ phases is that impurities accumulate at the racemic interfaces between the chiral domains with opposite chirality, as shown in Figure 14; in this case as a result of thermal decomposition in the $Iso^{[*]}$ phase at high temperatures. In the example shown in Figure 14a, the "membranes" enclosing the domains with opposite chirality are stable and the content retains uniform chirality in the liquid state. However, the "membranes" are not sufficiently stable to resist the chirality synchronization taking place across these boundaries at the phase transition to the cubic *I*23 phase. Therefore, the boundaries between the chiral domains are shifted, whereas the position of the walls between the compartments does not change (most probably due to their pinning to the surfaces), and thus the formed "membranes" become visible as thin lines in the Cub_{bi}/*I*23 phase (arrow in Figure 14b,c). This could be considered as another alternative way for the transition from local to total mirror symmetry breaking in the $Iso^{[*]}$ phase, in this case by isolating the domains with opposite chirality from each other, i.e., by reducing the length scale of the considered system, simply by enclosing and separating the different domains. It could be expected that the addition of properly chosen molecules, which are incompatible with the material forming the bulk $Iso^{[*]}$ phase, can lead to more stable membranes, making the compartments more resistant to external influences. This provides an interesting mode of symmetry breaking, combining local symmetry breaking with compartmentalization in a single unified process. This could be a possible mechanism of the development of uniform chirality in abiogenesis, as discussed further below.

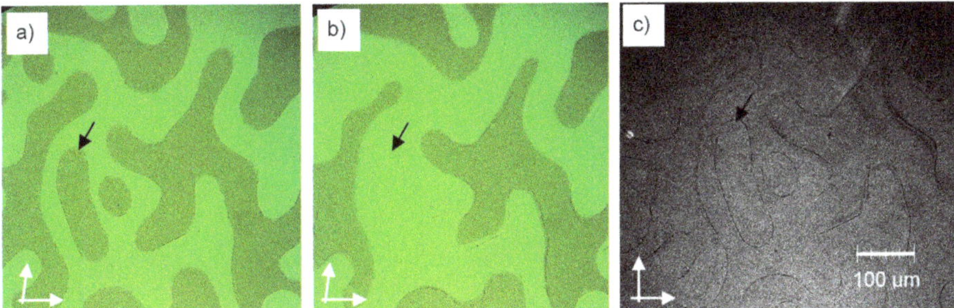

Figure 14. Compartmentalization by mirror symmetry breaking. (**a**) Chiral domains in the $Iso^{[*]}$ phase of compound **3** (*m*, *n* = 10) at *T* = 177 °C. After annealing at this temperature for about an hour thermal decomposition products accumulate at the racemic walls between the domains with opposite chirality; (**b**) shows the same range after transition to the Cub_{bi}/*I*23 phase upon cooling to *T* = 175 °C. Because of the slow seed formation, the borders between the chiral domains are shifted and the walls separating the chiral domains become visible. In this case, the walls are obviously not defect-free, meaning that they can still be passed by the chiral information. (**c**) shows (**b**) with crossed polarizers, and enlarges exposure time (previously unpublished results).

5.8. Impact of Network Formation on Mirror Symmetry Breaking in Isotropic Liquids

What is the origin of chirality in these liquids? These liquids occur, in most cases adjacent to one of the Cub_{bi} phases with helical network structure, either the achiral *Ia*3*d* or the mirror symmetry broken *I*23 phase. Therefore, it is likely that in the liquid state fragments of these, Cub_{bi} phases are also retained as local clusters, leading to a liquid polyamorphism [142,143]. As shown in Figure 15 for the example of the benzil derivative **10** (Figure 6, *n* = 12) with the $Iso^{[*]}$ phase occurring besides a *Ia*3*d* phase, the development of this cubic phase takes place from the isotropic liquid state by a series of phase transitions. These are indicated by three distinct exotherms in the DSC cooling traces, one dominating broad feature and two sharper transitions with smaller peaks [130] which can be interpreted as follows. With lowering temperature, nucleation of the molecules sets in, thus leading to clusters (cybotaxis), which increase in size. This clustering is supported by the nanoscale segregation of incompatible molecular parts, the rigid polyaromatic cores and the flexible alkyl chains [93,97,117], leading to short helix fragments. With further lowering temperature, the helical clusters grow and become interconnected to dynamic random networks (Figure 15b), which further grow and increase in connectivity in the range of the broad feature in the DSC trace (as introduced in Section 5.6, we designate this achiral liquid with a local network structure as Iso_1, see Figure 15a) [144]. At a certain critical connectivity, the helix sense is spontaneously synchronized over macroscopic distances, leading to the mirror symmetry broken liquid ($Iso^{[*]}$ phase) [25,130]. The relatively sharp peak for this Iso_1-$Iso^{[*]}$ transition (Figure 15a) is likely to be mainly the result of the chiral segregation at this transition [25]. Further increasing network connectivity finally leads to the phase transition to the Cub_{bi} phase, where a long-range positional order is established [40,130].

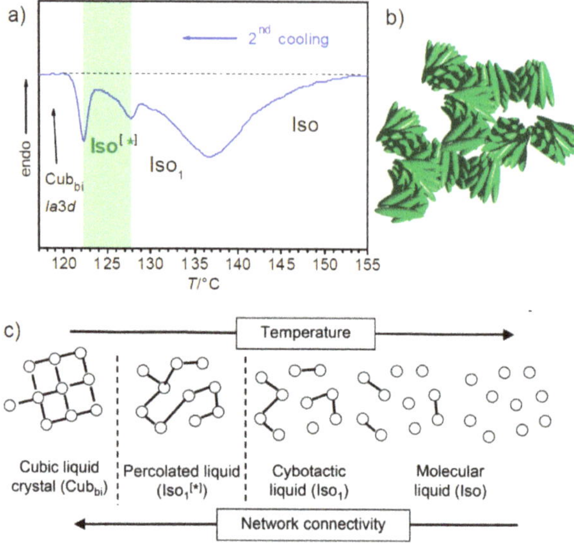

Figure 15. Investigation of the benzil compound **10** (Figure 6, $n = 12$). (**a**) DSC cooling trace (10 K min^{-1}) showing the transitions Iso-Iso$_1$-Iso$^{[*]}$-Cub$_{bi}$/$Ia3d$; (**b**) shows a model of the rod-like molecules in the dynamic helical networks of the Iso$^{[*]}$ phase and (**c**) shows schematically the transition from an ordinary liquid via cybotactic and percolated network liquids to the Cub$_{bi}$ phase with increasing network connectivity [40,42,130]; the dots symbolize the helical nano-clusters and the lines the interactions between them. It is noted that the Iso-Iso$_1$ transition cannot be identified in the DSC traces as a separate phase transition in all cases, either if this transition is completely continuous or if the aggregate structure of the liquid is present in the whole achiral Iso range. This case is found for compounds **3** with a diphenyl-dithiophene core unit [25]; reprinted from [130].

Depending on the type of formed cubic lattice in the resulting cubic phase, the chirality of the Iso$^{[*]}$ phases is either retained at the transition to the chiral $I23$ phase or removed at the transition to the achiral $Ia3d$ phase [26]. As the Iso$^{[*]}$-$Ia3d$ transition requires helix reversal for half of the molecules, this transition is slow and can be significantly supercooled compared to the transition on heating [137]. The whole sequence is shown schematically in Figure 15c, where the dots represent the local domains and the lines indicate their increasing connectivity to a network structure. Thus, the chirality synchronization and mirror symmetry breaking are supported by the presence of a dynamic network structure, even in the liquid state (percolated liquids, network liquids) [145–148]. The local structure of the networks in the Iso$^{[*]}$ phase could either be completely random or $I23$-like, but not $Ia3d$-like, which would retain an achiral liquid state. There is no visible change in the optical rotation at the Iso$^{[*]}$-$I23$ transition (Figures 13 and 14), and therefore we can assume that the local structure is $I23$-like also in the Iso$^{[*]}$ phase and that segregation of the helicity is complete in both phases, i.e., there should be uniform helix sense in the domains of these C-type systems. It is much more difficult to answer the question of the molecular helicity [25,40,42]. The compounds discussed here have no stereogenic centre, and hence are achiral. Nevertheless, these molecules can assume chiral conformations (see for example Figure 9a,b) [25,42]. Because the energy barrier between these conformers is relatively small, there is a conformational equilibrium of the enantiomorphic helices and this equilibrium is biased to some extent by the helical organization of the molecules towards one of the enantiomorphic conformers, leading to the energy minimum diastereomeric pair. Moreover, a denser packing is obviously achieved for conformers with uniform chirality and this effect is strengthened and long-range transmitted by the organization of the molecules in networks. Although we are sure about the homochirality of the helical networks, the degree of segregation of enantiomorphic conformers in these networks is more difficult to estimate and likely to be scalemic.

This raises the important question of if transient molecular chirality is required for helix formation, or if helix formation in liquids would be possible even without any contribution of molecular helicity. This is difficult to prove experimentally, because completely rigid polyaromatic units tend to crystallize instead of forming Iso$^{[*]}$ phases. Nevertheless, this work has shown that, in C-type systems, chirality can spontaneously develop from achiral molecules as thermodynamically stable phases, even in the liquid or liquid crystalline state.

6. Possible Scenario for Chirogenesis Based on Mirror Symmetry Breaking in Network Liquids

Though it is self-evident that the reported system cannot have played any role in abiogenesis, this observation shows that chirality can arise spontaneously by liquid state self-assembly, just like the Soai reaction, which confirmed the concept of autocatalysis for the emergence of uniform chirality in autocatalytic cycles of chemical reactions [57,58]. The chirality develops under thermodynamic control, meaning that it is long-term stable and can be retained without the requirement of complex catalytic cycles. It allows the development of spontaneous chirality in liquids, even at the temperatures of hydrothermal vents, where abiogenesis is assumed to have started [149,150]. As shown above, the Iso$^{[*]}$ phase is not restricted to local mirror symmetry breaking, but can also develop total mirror symmetry breaking either *(i)* stochastically by cycling across phase transitions or *(ii)* by compartmentalization, or *(iii)* deterministic under the influence of extremely weak chiral fields by chirality amplification. Toxvaerd proposed different chirogenesis scenarios of mirror symmetry breaking in fluids based on developing carbohydrate [19] and protein chirality [151]. However, we have shown that it works even with achiral molecules [25]. The only requirements are a C-type system, a rapid enantiomerization and network formation in a fluid system [26,40,129]. This raises the question of whether a related process of chirality synchronization in the liquid state could have been of importance for the emergence of homochirality during abiogenesis.

A possible hypothesis could be the following. If one believes the RNA world picture of the origin of life [152–154] and that chirality had to be selected before biogenesis, then homochirality perhaps developed together with the formation of early forms of RNA (proto-RNA or pre-RNA) [155,156]. As nowadays, prebiotically plausible ways towards nucleotides are known [4,154,157–159], the process could have started with the self-assembly of a library of these amphiphilic *N*-containing heteroaromatic compounds, forming (slightly twisted) hydrogen-bonded pairs or oligomers, stacking with their faces on top of each other (as still found in today's RNA and DNA). The dense packing expels the water from the core region, thus forming nano-segregated rods of helically twisted π-stacked *N*-heterocycles in the aqueous continuum (Figure 16a). With increasing concentrations of these short helices, they fuse to dynamic networks with long-range synchronized helicity, thus forming a mirror symmetry broken liquid (or a mirror symmetry broken Cub_{bi} liquid crystal at an even higher concentration, Figure 16b). The polar groups, together with excess water, form the continuum in a very similar way to the alkyl chains in the case of the previously discussed polycatenars compounds. This helical organization might not only be driven by the different size of aromatic units and polar surrounding (Figure 8c), but also by the well-known tendency of staggered packing of these flat, electron-poor *N*-heterocycles [160–162].

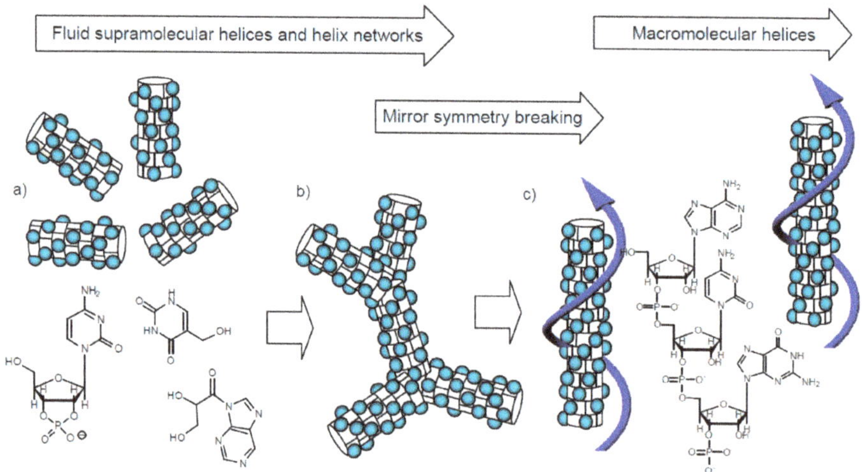

Figure 16. Hypothesis of development of biochirality: (**a**) Libraries of amphiphilic *N*-heterocycles self-assemble into dynamic helices (the blue dots represent the polar groups, being carbohydrates or other units) which (**b**) become chirality synchronized by dynamic network formation in the mirror symmetry broken helical network fluid and (**c**) polymerization leads to proto-RNA with fixed helicity and uniform absolute configuration (chirality) of the attached carbohydrates.

Thus, the helical self-assembly is determined by the achiral *N*-heterocyclic units, whereas the attached polar units are either achiral or (almost or completely) racemic and have no effect on the helix formation itself. The actual sense of helicity of these dynamic aggregates is either determined by local fluctuations (stochastic outcome) or by any kind of weak chiral field, leading to a deterministic outcome with preference for one sense of chirality. During growth of the dynamic helical networks, the building blocks were continuously exchanged to achieve the most stable helical system. This also includes the exchange of the wrong enantiomers. The formation of covalent bonds between properly preorganized groups in the polar periphery fixes the helical structure and leads to the development of dynamic proto-RNA oligomers (Figure 16c), in which the exchange and optimization continues due to the dynamics resulting from ongoing hydrolysis/condensation reactions as well as epimerization and racemization (transition from dynamic supramolecular aggregates to dynamic covalent bonds). The helix sense and also the once-chosen chirality of the carbohydrates or carbohydrate-like units (i.e., their relative and absolute configuration) become more and more fixed by the confinement provided by fixation in the developing polymer system [82,163–169]. Thus, the chirality synchronization is followed by covalent fixation. After reaching a critical polymer stability and after reaching a certain limiting polymer length, the interconnected network structure has to be given up, and from that point on, the helical proto-RNA can act as a relatively persistent carrier of the once-established chiral information (transition from dynamic to permanent covalent bonds). This would mean that uniform chirality is likely to have co-evolved during proto-RNA formation, which has fixed the carbohydrate chirality.

That chirality synchronization can indeed be observed in self-assembled soft matter systems formed by simple achiral amphiphilic *N*-heterocycles was recently shown experimentally for compound **11** (Figure 17) [170], where the oligoethylene glycol chain represents the polar group, replacing the ribose and phosphate units of the related nucleic acids. In the investigated case, it forms a glassy conglomerate of chiral domains in the bulk state (without solvent), as shown in Figure 17a,b [170]. The spontaneous formation of homochiral domains was recently also observed in supramolecular polymer gels formed in aqueous systems of self-assembled helical aggregates of hydrogen-bonded achiral heterocycles [171].

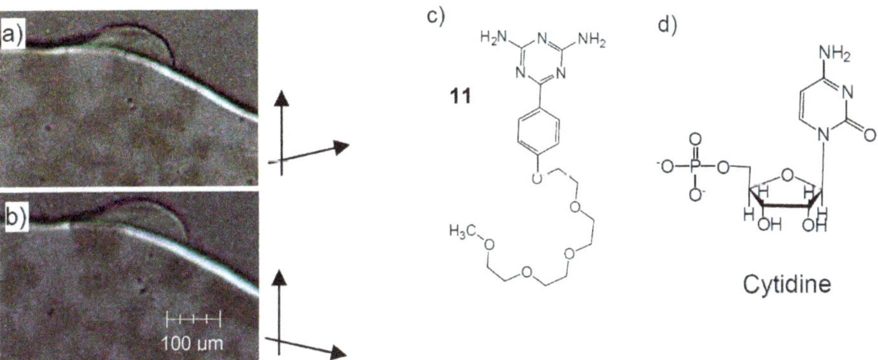

Figure 17. Amphiphilic *N*-heterocyclic compound **11** forming a glassy Iso[*] phase. (**a**,**b**) show the chiral domains as observed between slightly uncrossed polarizers at 90 °C, (**c**) shows the molecular structure and (**d**) shows the structure of a nucleotide for comparison, adapted from [170].

Though peptides might have co-evolved together with RNA [82], peptide self-assembly itself is less likely to be the source of homochirality, though peptides are capable of helix formation and hydrophobic amino acids could give rise to amphiphilicity and aggregation [151], but they are unlikely to be able to form mirror symmetry broken liquid states. The reason for this is that for flexible amphiphiles (lipids [96], block-copolymers [101]), only double networks have been found to date, usually the achiral double-gyroid *Ia*3d phase. However, mirror symmetry breaking in network fluids requires a chiral network structure, as provided by the *I*23-like triple network, which has, to date, only been found for molecules involving sufficiently large π-conjugated aromatic units and this can be provided by the hydrogen-bonded pairs (or larger aggregates) of nucleobases and related *N*-heterocycles. This does not exclude that *N*-heterocycles of aromatic amino acids could also have been involved in the chirogenesis of network fluids and transferred their chirality to the developing peptides. This would be in line with a co-evolution scenario [81].

It is also likely that this chirogenesis process, which requires relatively high concentrations to retain the chirality synchronized fluid state, took place parallel in confined spaces like pores and with a different outcome in different pores. In larger spaces, the possibility of compartmentalization of the fluid conglomerates by developing walls formed by lipids and other lipophilic molecules accumulating at the racemic boundaries between the aqueous domains with opposite chirality would allow the coupled formation of protocells and uniform chirality. In both cases, the chirality is only selected within the individual compartments, and the overall outcome could still be racemic (stochastic), but one of the enantiomeric cell types can then become dominant in the following selection processes [5]. However, the huge chirality amplification power of the mirror symmetry broken network fluids [25] makes it more likely that the sign of chirality has developed in a more deterministic way under the influence of weak chirality fields (meteorites, chiral surfaces, circular polarized light, etc.) [9]. Once the dominating chirality sense is fixed, the weak chiral fields cannot alter the once selected chirality.

Thus, the early RNA could have not only acted as the first information carrier and catalyst, it might also have provided uniform biochirality. Accordingly, biochirality would have developed from *achiral* units, capable of forming dynamic helical networks by self-assembly into a chiral prebiotic fluid, and this triggered the formation of the biorelevant molecules which were homogeneously chiral (or at least scalemic) right from the beginning (see Figure 18). Thus, chirality can probably be considered as the first information stored by the emerging RNA. This model is only based on thermodynamic arguments, molecular self-assembly and phase transitions. It does not require complex chemical reaction networks, kinetically stabilized stationary states and a continuous flow of energy or material, and therefore is likely to be compatible with the situation at the very beginning of abiogenesis.

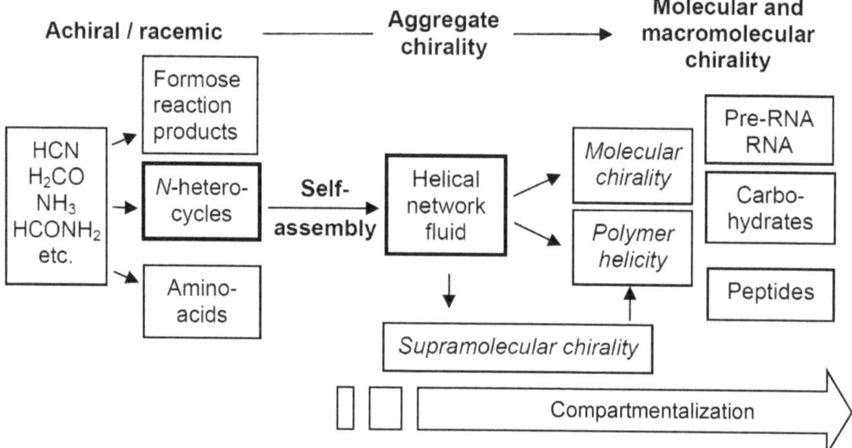

Figure 18. Scheme showing a possible transition from simple achiral to chiral molecules and complex functional biopolymers in abiogenesis, being mediated by chiral network fluids.

Overall, during abiogenesis, the complexity of chemical systems has increased by a combination of chemical reactions and self-assembly (molecular and supramolecular self-assembly). Abiogenesis presumably started with small achiral molecules (HCN, NH_3, H_2CO, etc.) where chirality did not play a role. They condensed to larger molecules (heterocycles, formose reaction products and amino acids) where chirality became an issue. Some of these molecules were capable of soft helical self-assembly, forming fluid networks capable of chirality synchronization. The developing uniform chirality then allowed the formation of polymers and the fixation of the selected chirality sense, and homochiral molecules and polymers were formed. The emerging molecular homochirality then triggered the transition from simple chemistry to complex reaction networks and catalytic and autocatalytic metabolic cycles, which gave rise to non-equilibrium chemistry in abiogenesis (transition from self-assembly to self-organization), finally leading to the Darwinistic evolution of life in the continuing biogenesis.

7. Conclusions

Recent progress in mirror symmetry breaking and chirality amplification in isotropic liquids and liquid crystalline cubic phases of achiral molecule is reviewed and discussed in relation to the autocatalytic Soai chirality amplification reaction, the Viedma type deracemizations, and with respect to its implications for the hypothesis of emergence of biological chirality. It is shown that this symmetry breaking process requires fluid systems where homochiral interactions are preferred over heterochiral (C-type systems). Moreover, it is experimentally shown that dynamic network structures lead to chirality synchronization if the enantiomerization barrier is sufficiently low, i.e., that racemization drives the development of uniform chirality. The typical outcome is conglomerate formation, indicating a local mirror symmetry breaking. This led to the first experimentally proven example of phase separation in an isotropic liquid solely based on mirror symmetry breaking. Total mirror symmetry breaking was found by crossing phase transitions due to kinetic effects, or under the influence of minor chiral fields, leading to stochastic and deterministic homochirality, respectively. This is associated with the extreme chirality amplification power of these systems, especially close to the bifurcation point. Once formed, these mirror symmetry broken liquids are longtime stable because they represent thermodynamic minimum systems. This mode of spontaneous mirror symmetry breaking in the isotropic liquid state is considered as important for the emergence of biochirality. A model is hypothesized, which assumes the emergence of uniform chirality by dynamic helical networks in fluids formed by achiral heterocycles in water, followed by polymerization, which fixes chirality and simultaneously leads to (proto)-RNA formation and uniform biochirality. Possibly, this process was even combined with a liquid–liquid compartmentalization, providing a dynamic cellular structure.

Funding: Parts of this work were funded by the Deutsche Forschungsgemeinschaft, grant number Ts 39/24-2.

Conflicts of Interest: The authors declare no conflict of interest. The funders had no role in the design of the study; in the collection, analyses, or interpretation of data; in the writing of the manuscript, or in the decision to publish the results.

References

1. Pasteur, L. Recherches sur les relations qui peuvent exister entre la forme crystalline, la composition chimique et le sens de la polarisation rotatoire. *Ann. Chim. Phys.* **1848**, *24*, 442–459.
2. Palyi, G. *Biological Chirality*; Academic Press, Elsevier: London, UK, 2020.
3. Weiss, M.C.; Preiner, M.; Xavier, J.C.; Zimorski, V.; Martin, W.F. The last universal common ancestor between ancient Earth chemistry and the onset of Genetics. *PLoS Genet.* **2018**, *14*, e1007518. [CrossRef] [PubMed]
4. Islam, S.; Powner, M.W. Prebiotic Systems Chemistry: Complexity Overcoming Clutter. *Chem* **2017**, *2*, 470–501. [CrossRef]
5. Green, M.M.; Jain, V. Homochirality in Life: Two Equal Runners, One Tripped. *Orig. Life Evol. Biosph.* **2010**, *40*, 111–118. [CrossRef] [PubMed]
6. Walker, S.I. Origins of life: A problem for physics, a key issues review. *Rep. Prog. Phys.* **2017**, *80*, 092601. [CrossRef] [PubMed]
7. Ruiz-Mirazo, K.; Briones, C.; de la Escosura, A. Prebiotic Systems Chemistry: New Perspectives for the Origins of Life. *Chem. Rev.* **2014**, *114*, 285–366. [CrossRef] [PubMed]
8. Rauchfuss, H. *Chemical Evolution and the Origin of Life*; Springer: Berlin/Heidelberg, Germany, 2008.
9. Guijarro, A.; Yus, M. *The Origin of Chirality in the Molecules of Life*; RSC publishing: Cambridge, UK, 2009.
10. Luisi, P.L. *The Emergence of Life*; Cambridge University Press: Cambridge, UK, 2006.
11. Meierhenrich, U. *Amino Acids and the Asymmetry of Life*; Springer: Berlin/Heidelberg, Germany, 2008.
12. Cintas, P. *Biochirality, Origins, Evolution and Molecular Recognition*; Springer: Berlin/Heidelberg, Germany, 2013.
13. Cronin, J.; Reisse, J. Chirality and the Origin of Homochirality. In *Lectures in Astrobiology*; Gargaud, M., Barbier, B., Martin, H., Reise, J., Eds.; Springer: Berlin/Heidelberg, Germany, 2005; Volume 1, pp. 473–515.
14. Blackmond, D.G. The Origin of Biological Homochirality. *Cold Spring Harb. Perspect. Biol.* **2019**, *11*, a032540. [CrossRef] [PubMed]
15. Coveney, P.V.; Swadling, J.B.; Wattis, J.A.D.; Greenwell, H.C. Theory, modelling and simulation in origins of life studies. *Chem. Soc. Rev.* **2012**, *41*, 5430–5446. [CrossRef] [PubMed]
16. Ribo, J.M.; Hochberg, D.; Crusats, J.; El-Hachemi, Z.; Moyano, A. Spontaneous mirror symmetry breaking and origin of biological homochirality. *J. R. Soc. Interface* **2017**, *14*, 20170699. [CrossRef] [PubMed]
17. Ribó, J.M.; Hochberg, D. Chemical Basis of Biological Homochirality during the Abiotic Evolution Stages on Earth. *Symmetry* **2019**, *11*, 814. [CrossRef]
18. Pavlov, V.A.; Shushenachev, Y.V.; Zlotin, S.G. Chiral and Racemic Fields Concept for Understanding of the Homochirality Origin, Asymmetric Catalysis, Chiral Superstructure Formation from Achiral Molecules, and B-Z DNA Conformational Transition. *Symmetry* **2019**, *11*, 649. [CrossRef]
19. Toxvaerd, S. The Role of Carbohydrates at the Origin of Homochirality in Biosystems. *Orig. Life Evol. Biosph.* **2013**, *43*, 391–409. [CrossRef] [PubMed]
20. Toxvaerd, S. A Prerequisite for Life. *J. Theor. Biol.* **2019**, *474*, 48–51. [CrossRef] [PubMed]
21. Hein, J.E.; Blackmond, D.G. On the Origin of Single Chirality of Amino Acids and Sugars in Biogenesis. *Acc. Chem. Res.* **2012**, *45*, 2045–2054. [CrossRef]
22. Pavlov, V.A.; Klabunovskii, E.I. Homochirality Origin in Nature: Possible Versions. *Curr. Org. Chem.* **2014**, *18*, 93–114. [CrossRef]

23. Avalos, M.; Babiano, R.; Cintas, P.; Jimenez, J.L.; Palacios, J.C. Homochirality and chemical evolution: New vistas and reflections on recent models. *Tetrahedron Asymmetry* **2010**, *21*, 1030–1040. [CrossRef]
24. Suzuki, N.; Itabashi, Y. Possible Roles of Amphiphilic Molecules in the Origin of Biological Homochirality. *Symmetry* **2019**, *11*, 966. [CrossRef]
25. Dressel, C.; Reppe, T.; Prehm, M.; Brautzsch, M.; Tschierske, C. Chiral self-sorting and amplification in isotropic liquids of achiral molecules. *Nat. Chem.* **2014**, *6*, 971–977. [CrossRef]
26. Dressel, C.; Liu, F.; Prehm, M.; Zeng, X.B.; Ungar, G.; Tschierske, C. Dynamic Mirror-Symmetry Breaking in Bicontinuous Cubic Phases. *Angew. Chem. Int. Ed.* **2014**, *53*, 13115–13120. [CrossRef]
27. Yashima, E.; Ousaka, N.; Taura, D.; Shimomura, K.; Ikai, T.; Maeda, K. Supramolecular helical systems: Helical assemblies of small molecules, foldamers, and polymers with chiral amplification and their functions. *Chem. Rev.* **2016**, *116*, 13752–13990. [CrossRef]
28. Liu, M.; Zhang, L.; Wang, T. Supramolecular chirality in self-assembled systems. *Chem. Rev.* **2015**, *115*, 7304–7397. [CrossRef]
29. Barclay, T.G.; Constantopoulos, K.; Matisons, J. Nanotubes self-assembled from amphiphilic molecules via helical intermediates. *Chem. Rev.* **2014**, *114*, 10217–10291. [CrossRef]
30. La, A.D.D.; Al Kobaisi, M.; Gupta, A.; Bhosale, S.V. Chiral Assembly of AIE-Active Achiral Molecules: An Odd Effect in Self-Assembly. *Chem. Eur. J.* **2017**, *23*, 3950–3956. [CrossRef]
31. Sang, Y.; Liu, M. Symmetry Breaking in Self-Assembled Nanoassemblies. *Symmetry* **2019**, *11*, 950. [CrossRef]
32. Changa, B.; Lib, X.; Suna, T. Self-assembled chiral materials from achiral components or racemates. *Eur. Polym. J.* **2019**, *118*, 365–381. [CrossRef]
33. Hoeben, F.J.M.; Jonkheijm, P.; Meijer, E.W.; Schenning, A.P.H.J. About Supramolecular Assemblies of π-Conjugated Systems. *Chem. Rev.* **2005**, *105*, 1491–1546. [CrossRef]
34. Palmans, A.R.A.; Meijer, E.W. Amplification of chirality in dynamic supramolecular Aggregates. *Angew. Chem. Int. Ed.* **2007**, *46*, 8948–8968. [CrossRef]
35. Zhang, L.; Wang, T.; Shen, Z.; Liu, M. Chiral Nanoarchitectonics: Towards the Design, Self-Assembly, and Function of Nanoscale Chiral Twists and Helices. *Adv. Mater.* **2016**, *28*, 1044–1059. [CrossRef]
36. Pijper, D.; Feringa, B.L. Control of dynamic helicity at the macro- and supramolecular level. *Soft Matter* **2008**, *4*, 1349–1372. [CrossRef]
37. Ariga, K.; Mori, T.; Kitao, T.; Uemura, T. Supramolecular Chiral Nanoarchitectonics. *Adv. Mater.* **2020**, 1905657. [CrossRef]
38. De Greef, T.F.A.; Smulders, M.M.J.; Wolffs, M.; Schenning, A.P.H.J.; Sijbesma, R.P.; Meijer, E.W. Supramolecular polymerization. *Chem. Rev.* **2009**, *109*, 5687–5754. [CrossRef] [PubMed]
39. Lehmann, A.; Alaasar, M.; Poppe, M.; Poppe, S.; Prehm, M.; Nagaraj, M.; Sreenilayam, S.P.; Panarin, J.P.; Vij, J.K.; Tschierske, C. Stereochemical Rules Govern the Soft Self-Assembly of Achiral Compounds: Understanding the Heliconical Liquid-Crystalline Phases of Bent-Core Mesogens. *Chem. Eur. J.* **2020**, *26*, 4714–4733. [CrossRef] [PubMed]
40. Tschierske, C. Mirror symmetry breaking in liquids and liquid crystals. *Liq. Cryst.* **2018**, *45*, 2221–2252. [CrossRef]
41. Le, K.V.; Takezoe, H.; Araoka, F. Chiral Superstructure mesophases of achiral bent-shaped molecules-hierarchical chirality amplification and physical properties. *Adv. Mater.* **2017**, *29*, 1602737. [CrossRef] [PubMed]
42. Tschierske, C.; Ungar, G. Mirror Symmetry Breaking by Chirality Synchronisation in Liquids and Liquid Crystals of Achiral Molecules. *ChemPhysChem* **2016**, *17*, 9–26. [CrossRef]
43. Dierking, I. Chiral Liquid Crystals: Structures, Phases, Effects. *Symmetry* **2014**, *6*, 444–472. [CrossRef]
44. Nishyiama, I. Remarkable Effect of Pre-organization on the Self Assembly in Chiral Liquid Crystals. *Chem. Rec.* **2010**, *9*, 340–355. [CrossRef] [PubMed]
45. Bisoyi, H.K.; Bunning, T.J.; Li, Q. Stimuli-Driven Control of the Helical Axis of Self-Organized Soft Helical Superstructures. *Adv. Mater.* **2018**, *30*, 1706512. [CrossRef]
46. Eliel, E.L.; Wilen, S.H.; Mander, L.N. *Stereochemistry of Organic Compounds*; Wiley: New York, NY, USA, 1994.
47. Jacques, J.; Collet, A.; Wilen, S.H. *Enantiomers, Racemates and Resolutions*; Wiley: New York, NY, USA, 1981.
48. Mills, W.H. Some Aspects of Stereochemistry. *Chem. Ind.* **1932**, *51*, 750–759. [CrossRef]
49. Morowitz, M. A mechanism for the amplification of fluctuations in racemic mixtures. *J. Theor. Biol.* **1969**, *25*, 491–494. [CrossRef]

50. Siegel, J. Homochiral Imperative of Molecular Evolution. *Chirality* **1998**, *10*, 24–27. [CrossRef]
51. Lente, G. The Role of Stochastic Models in Interpreting the Origins of Biological Chirality. *Symmetry* **2010**, *2*, 767–798. [CrossRef]
52. Hochberg, D.; Zorzano, M.-P. Reaction-noise induced homochirality. *Chem. Phys. Lett.* **2006**, *431*, 185–189. [CrossRef]
53. Kondepudi, D.K.; Aaskura, K. Chiral Autocatalysis, Spontaneous Symmetry Breaking, and Stochastic Behavior. *Acc. Chem. Res.* **2001**, *34*, 946–954. [CrossRef]
54. Silva-Dias, L.; Lopez-Castillo, A. Stochastic chiral symmetry breaking process besides the deterministic one. *Phys. Chem. Chem. Phys.* **2017**, *19*, 29424–29428. [CrossRef] [PubMed]
55. Frank, F.C. On spontaneous asymmetric synthesis. *Biochim. Biophys. Acta* **1953**, *11*, 459–464. [CrossRef]
56. Weissbuch, I.; Leiserowitz, L.; Lahav, M. Stochastic "Mirror Symmetry Breaking" via Self-Assembly, Reactivity and Amplification of Chirality: Relevance to Abiotic Conditions. *Top. Curr. Chem.* **2005**, *259*, 123–165. [CrossRef]
57. Soai, K.; Kawasaki, T.; Matsumoto, A. The Origins of Homochirality Examined by Using Asymmetric Autocatalysis. *Chem. Rec.* **2014**, *14*, 70–83. [CrossRef]
58. Soai, K.; Kawasaki, T.; Matsumoto, A. Role of Asymmetric Autocatalysis in the Elucidation of Origins of Homochirality of Organic Compounds. *Symmetry* **2019**, *11*, 694. [CrossRef]
59. Satyanarayana, T.; Abraham, S.; Kagan, H.B. Nonlinear Effects in Asymmetric Catalysis. *Angew. Chem. Int. Ed.* **2009**, *48*, 456–494. [CrossRef]
60. Kawasaki, T.; Tanaka, H.; Tsutsumi, T.; Kasahara, T.; Sato, I.; Soai, K. Chiral Discrimination of Cryptochiral Saturated Quaternary and Tertiary Hydrocarbons by Asymmetric Autocatalysis. *J. Am. Chem. Soc.* **2006**, *128*, 6032–6033. [CrossRef] [PubMed]
61. Kawasaki, T.; Okano, Y.; Suzuki, E.; Takano, S.; Oji, S.; Soai, K. Asymmetric Autocatalysis: Triggered by Chiral Isotopomer Arising from Oxygen Isotope Substitution. *Angew. Chem. Int. Ed.* **2011**, *50*, 8131–8133. [CrossRef] [PubMed]
62. Matsumoto, A.; Kaimori, Y.; Uchida, M.; Omori, H.; Kawasaki, T.; Soai, K. Achiral Inorganic Gypsum Acts as an Origin of Chirality through Its Enantiotopic Surface in Conjunction with Asymmetric Autocatalysis. *Angew. Chem. Int. Ed.* **2017**, *56*, 545–548. [CrossRef]
63. Soai, K.; Sato, I.; Shibata, T.; Komiya, S.; Hayashi, M.; Matsueda, Y.; Imamura, H.; Hayase, T.; Morioka, H.; Tabira, H.; et al. Asymmetric synthesis of pyrimidyl alkanol without adding chiral substances by the addition of diisopropylzinc to pyrimidine-5-carbaldehyde in conjunction with asymmetric autocatalysis. *Tetrahedron Asymmetry* **2003**, *14*, 185–188. [CrossRef]
64. Singleton, D.A.; Vo, L.K. A Few Molecules Can Control the Enantiomeric Outcome. Evidence Supporting Absolute Asymmetric Synthesis Using the Soai Asymmetric Autocatalysis. *Org. Lett.* **2003**, *5*, 4337–4339. [CrossRef]
65. Toxvaerd, S. Origin of homochirallity in biosystems. *Int. J. Mol. Sci.* **2009**, *10*, 1290–1299. [CrossRef] [PubMed]
66. Plasson, R.; Kondepudi, D.K.; Bersine, H.; Commeyras, A.; Asakura, K. Emergence of Homochirality in Far-From-Equilibrium Systems: Mechanisms and Role in Prebiotic Chemistry. *Chirality* **2007**, *19*, 589–600. [CrossRef] [PubMed]
67. Ribo, J.M.; Blanco, C.; Crusats, J.; El-Hachemi, Z.; Hochberg, D.; Moyano, A. Absolute Asymmetric Synthesis in Enantioselective Autocatalytic Reaction Networks: Theoretical Games, Speculations on Chemical Evolution and Perhaps a Synthetic Option. *Chem. Eur. J.* **2014**, *20*, 17250–17271. [CrossRef]
68. Toxvaerd, S. Molecular Dynamics Simulations in Isomerization Kinetics in Condensed Fluids. *Phys. Rev. Lett.* **2000**, *85*, 4747–4750.
69. Jafarpour, F.; Biancalani, T.; Goldenfeld, N. Noise-induced symmetry breaking far from equilibrium and the emergence of biological homochirality. *Phys. Rev. E* **2017**, *95*, 032407. [CrossRef]
70. Sugimori, T.; Hyuga, H.; Saito, Y. Fluctuation Induced Homochirality. *J. Phys. Soc. Jpn.* **2008**, *77*, 064606. [CrossRef]
71. Latinwo, F.; Stillinger, F.H.; Debenedetti, P.G. Molecular model for chirality phenomena. *J. Chem. Phys.* **2016**, *145*, 154503. [CrossRef]
72. Brandenburg, A. The Limited Roles of Autocatalysis and Enantiomeric Cross-Inhibition in Achieving Homochirality in Dilute Systems. *Orig. Life Evol. Biosph.* **2019**, *49*, 49–60. [CrossRef]

73. Kondepudi, D.K.; Kaufman, R.J.; Singh, N. Chiral Symmetry Breaking in Sodium Chlorate Crystallizaton. *Science* **1990**, *250*, 975–976. [CrossRef] [PubMed]
74. Viedma, C. Chiral Symmetry Breaking During Crystallization: Complete Chiral PurityInduced by Nonlinear Autocatalysis and Recycling. *Phys. Rev. Lett.* **2005**, *94*, 065504. [CrossRef] [PubMed]
75. Cintas, P.; Viedma, C. On the Physical Basis of Asymmetry and Homochirality. *Chirality* **2012**, *24*, 894–898. [CrossRef] [PubMed]
76. Blackmond, D.G. "Chiral Amnesia" as a Driving Force for Solid-Phase Homochirality. *Chem. Eur. J.* **2007**, *13*, 3290–3295. [CrossRef]
77. Noorduin, W.L.; Vlieg, E.; Kellogg, R.M.; Kaptein, B. From Ostwald Ripening to Single Chirality. *Angew. Chem. Int. Ed.* **2009**, *48*, 9600–9606. [CrossRef]
78. Amabilino, D.B.; Kellogg, R.M. Spontaneous Deracemization. *Isr. J. Chem.* **2011**, *51*, 1034–1040. [CrossRef]
79. Uemura, N.; Sano, K.; Matsumoto, A.; Yoshida, Y.; Mino, T.; Sakamoto, M. Absolute Asymmetric Synthesis of an Aspartic Acid Derivative from Prochiral Maleic Acid and Pyridine under Achiral Conditions. *Chem. Asian J.* **2019**, *14*, 4150–4153. [CrossRef]
80. Tsogoeva, S.B.; Wei, S.; Freund, M.; Mauksch, M. Generation of Highly Enantioenriched Crystalline Products in Reversible Asymmetric Reactions with Racemic or Achiral Catalysts. *Angew. Chem. Int. Ed.* **2009**, *48*, 590–594. [CrossRef] [PubMed]
81. McLaughlin, D.T.; Thao Nguyen, T.P.; Mengnjo, L.; Bian, C.; Leung, Y.H.; Goodfellow, E.; Ramrup, P.; Woo, S.; Cuccia, L.A. Viedma Ripening of Conglomerate Crystals of Achiral Molecules Monitored Using Solid-State Circular Dichroism. *Cryst. Growth Des.* **2014**, *14*, 1067–1076. [CrossRef]
82. Runnels, C.M.; Lanier, K.A.; Williams, J.K.; Bowman, J.C.; Petrov, A.S.; Hud, N.V.; Williams, L.D. Folding, Assembly, and Persistence: The Essential Nature and Origins of Biopolymers. *J. Mol. Evol.* **2018**, *86*, 598–610. [CrossRef] [PubMed]
83. Kenney, J.F.; Deiters, U.K. The evolution of multicomponent systems at high pressures Part IV. The genesis of optical activity in high-density, abiotic Ñuids. *Phys. Chem. Chem. Phys.* **2000**, *2*, 3163–3174. [CrossRef]
84. Hochberg, D.; Cintas, P. Does Pressure Break Mirror-Image Symmetry? A Perspective and New Insights. *ChemPhysChem* **2020**, *21*, 633–642. [CrossRef] [PubMed]
85. Leporl, L.; Mengherl, M.; Molllca, V. Discriminating Interactions between Chiral Molecules in the Liquid Phase: Effect on Volumetric Properties. *J. Phys. Chem.* **1983**, *87*, 3520–3525. [CrossRef]
86. Atik, Z.; Ewing, M.B.; McGlashan, M.L. Chiral Discrimination in Liquids. Excess Molar Volumes of (1 − x)A+ + xA, Where A Denotes Limonene, Fenchone, and a-Methylbenzylamine. *J. Phys. Chem.* **1981**, *85*, 3300–3303. [CrossRef]
87. Dressel, C.; Weissflog, W.; Tschierske, C. Spontaneous mirror symmetry breaking in a re-entrant isotropic liquid. *Chem. Commun.* **2015**, *51*, 15850–15853. [CrossRef]
88. Alaasar, M.; Poppe, S.; Dong, Q.; Liu, F.; Tschierske, C. Mirror symmetry breaking in cubic phases and isotropic liquids driven by hydrogen bonding. *Chem. Commun.* **2016**, *52*, 13869–13872. [CrossRef]
89. Alaasar, M.; Prehm, M.; Cao, Y.; Liu, F.; Tschierske, C. Spontaneous Mirror-Symmetry Breaking in Isotropic Liquid Phases of Photoisomerizable Achiral Molecules. *Angew. Chem. Int. Ed.* **2016**, *55*, 312–316. [CrossRef] [PubMed]
90. Alaasar, M.; Poppe, S.; Dong, Q.; Liu, F.; Tschierske, C. Isothermal Chirality Switching in Liquid-Crystalline Azobenzene Compounds with Non Polarized Light. *Angew. Chem. Int. Ed.* **2017**, *56*, 10801–10805. [CrossRef]
91. Goodby, J.W.; Collings, P.J.; Kato, T.; Tschierske, C.; Gleeson, H.F.; Raynes, P. *Handbook of Liquid Crystals*, 2nd ed; Wiley-VHC: Weinheim, Germany, 2014.
92. Kato, T.; Uchida, J.; Ichikawa, T.; Sakamoto, T. Functional Liquid Crystals towards the Next Generation of Materials. *Angew. Chem. Int. Ed.* **2018**, *57*, 4355–4371. [CrossRef]
93. Tschierske, C. Development of Structural Complexity by Liquid-Crystal Self-assembly. *Angew. Chem. Int. Ed.* **2013**, *52*, 8828–8878. [CrossRef]
94. Mitov, M. Cholesteric liquid crystals in living matter. *Soft Matter* **2017**, *13*, 4176–4209. [CrossRef] [PubMed]
95. Lynch, M.L.; Spicer, P.T. (Eds.) *Bicontinuous Liquid Crystals*; Surfactant Science Series 127; CRC Press—Taylor & Francis Group: Boca Raton, FL, USA, 2005.
96. Seddon, J.M.; Templer, R.H. Polymorphism of Lipid-Water Systems. In *Handbook of Biological Physics*; Lipowsky, R., Sackmann, E., Eds.; Elsevier: Amsterdam, the Netherlands, 1995; Volume 1, pp. 97–160.

97. Tschierske, C. Micro-segregation, molecular shape and molecular topology—Partners for the design of liquid crystalline materials with complex mesophase morphologies. *J. Mater. Chem.* **2001**, *11*, 2647–2671. [CrossRef]
98. Borisch, K.; Diele, S.; Göring, P.; Kresse, H.; Tschierske, C. Tailoring Thermotropic Cubic Mesophases: Amphiphilic Polyhydroxy Derivatives. *J. Mater. Chem.* **1998**, *8*, 529–543. [CrossRef]
99. Kutsumizu, S. Recent Progress in the Synthesis and Structural Clarification of Thermotropic Cubic Phases. *Isr. J. Chem.* **2012**, *52*, 844–853. [CrossRef]
100. Ungar, G.; Liu, F.; Zeng, X. Cubic and 3D Thermotropic Liquid Crystal Phases and Quasicrystals. In *Handbook of Liquid Crystals*, 2nd ed; Goodby, J.W., Collings, P.J., Kato, T., Tschierske, C., Gleeson, H.F., Raynes, P., Eds.; Wiley-VHC: Weinheim, Germany, 2014; Volume 5, pp. 363–436.
101. Meuler, A.J.; Hillmyer, M.A.; Bates, F.S. Ordered Network Mesostructures in Block Polymer Materials. *Macromolecules* **2009**, *42*, 7221–7250. [CrossRef]
102. Hyde, S.; Andersson, S.; Larsson, K.; Blum, Z.; Landh, T.; Lidin, S.; Ninham, B.W. *The Language of Shape, The Role of Curvature in Condensed Matter: Physics, Chemistry and Biology*; Elsevier: Amsterdam, The Netherlands, 1997.
103. Deamer, D. Liquid crystalline nanostructures: Organizing matrices for non-enzymatic nucleic acid polymerization. *Chem. Soc. Rev.* **2012**, *41*, 5375–5379. [CrossRef] [PubMed]
104. Hamley, I.W. Liquid crystal phase formation by biopolymers. *Soft Matter* **2010**, *6*, 1863–1871. [CrossRef]
105. Jewell, S.A. Living systems and liquid crystals. *Liq. Cryst.* **2011**, *38*, 1699–1714. [CrossRef]
106. Steward, G.T. Liquid crystals in biology II. Origins and processes of life. *Liq. Cryst.* **2004**, *31*, 443–471. [CrossRef]
107. Rey, A.D. Liquid crystal models of biological materials and processes. *Soft Matter* **2010**, *6*, 3402–3429. [CrossRef]
108. Giraud-Guille, M.M.; Belamie, E.; Mosser, G.; Helary, C.; Gobeaux, F.; Vigier, S. Liquid crystalline properties of type I collagen: Perspectives in tissue morphogenesis. *Comptes Rendus Chim.* **2008**, *11*, 245–252. [CrossRef]
109. Lydon, J. Microtubules: Nature's smartest mesogens—A liquid crystal model for cell division. *Liq. Cryst. Today* **2006**, *15*, 1–10. [CrossRef]
110. Bouligand, Y. Geometry and topology of cell membranes. In *Geometry in Condensed Matter Physics*; Sadoc, J.F., Ed.; World Scientific: Singapore, 1990; pp. 193–231.
111. Kulkarni, C.V. Lipid crystallization: From self-assembly to hierarchical and biological ordering. *Nanoscale* **2012**, *4*, 5779–5791. [CrossRef]
112. Livolant, F.; Levelut, A.M.; Doucet, J.; Benoit, J.P. The highly concentrated liquid crystalline phase of DNA is columnar hexagonal. *Nature* **1989**, *339*, 724–726. [CrossRef]
113. Leforestie, A.; Bertin, A.; Dubochet, J.; Richter, K.; Sartori Blanc, N.; Livolant, F. Expression of chirality in columnar hexagonal phases of DNA and nucleosomes. *Comptes Rendus Chim.* **2008**, *11*, 229–244. [CrossRef]
114. Leal, C.; Ewert, K.K.; Bouxsein, N.F.; Shirazi, R.S.; Lic, Y.; Safinya, C.R. Stacking of short DNA induces the gyroid cubic-to inverted hexagonal phase transition in lipid–DNA complexes. *Soft Matter* **2013**, *9*, 795–804. [CrossRef]
115. Zanchetta, G. Spontaneous self-assembly of nucleic acids: Liquid crystal condensation of complementary sequences in mixtures of DNA and RNA oligomers. *Liq. Cryst. Today* **2009**, *18*, 40–49. [CrossRef]
116. Mahadevi, S.A.; Sastry, G.N. Cooperativity in Noncovalent Interactions. *Chem. Rev.* **2016**, *116*, 2775–2825. [CrossRef] [PubMed]
117. Tschierske, C. Microsegregation: From Basic Concepts to Complexity in Liquid Crystal Self-Assembly. *Isr. J. Chem.* **2012**, *52*, 935–959. [CrossRef]
118. Rest, C.; Kandanelli, R.; Fernandez, G. Strategies to create hierarchical self-assembled structures via cooperative non-covalent interactions. *Chem. Soc. Rev.* **2015**, *44*, 2543–2572. [CrossRef]
119. Takezoe, H. Spontaneous Achiral Symmetry Breaking in Liquid Crystalline Phases. *Top. Curr. Chem.* **2012**, *318*, 303–330. [CrossRef] [PubMed]
120. Chen, C.; Kieffer, R.; Ebert, H.; Prehm, M.; Zhang, R.-B.; Zeng, X.; Liu, F.; Ungar, G.; Tschierske, C. Chirality Induction through Nano-Phase Separation: Alternating Network Gyroid Phase by Thermotropic Self-Assembly of X-Shaped Bolapolyphiles. *Angew. Chem. Int. Ed.* **2020**, *59*, 2725–2729. [CrossRef] [PubMed]
121. Zeng, X.; Poppe, S.; Lehmann, A.; Prehm, M.; Chen, C.; Liu, F.; Lu, H.; Ungar, G.; Tschierske, C. A Self-assembled Bicontinuous Cubic Phase with Single Diamond Network. *Angew. Chem. Int. Ed.* **2019**, *131*, 7453–7457. [CrossRef]

122. Poppe, S.; Cheng, X.; Chen, C.; Zeng, X.; Zhang, R.-B.; Liu, F.; Ungar, G.; Tschierske, C. Liquid Organic Frameworks: The Single-Network "Plumber's Nightmare" Bicontinuous Cubic Liquid Crystal. *J. Am. Chem. Soc.* **2020**, *142*, 3296–3300. [CrossRef]
123. Zeng, X.B.; Ungar, G. Spontaneously chiral cubic liquid crystal: Three interpenetrating networks with a twist. *J. Mater. Chem. C* **2020**, *8*, 5389–5398. [CrossRef]
124. Kauffman, S. *At Home in the Universe*; Oxford University Press: Oxford, UK, 1995.
125. Malthete, J.; Nguyen, H.T.; Destrade, C. Phasmids and polycatenar mesogens. *Liq. Cryst.* **1993**, *13*, 171–187. [CrossRef]
126. Nguyen, H.T.; Destrade, C.; Malthete, J. Phasmids and Polycatenar Mesogens. *Adv. Mater.* **1997**, *9*, 375–388. [CrossRef]
127. Bruce, D.W. Calamitics, Cubics, and Columnars. Liquid-Crystalline Complexes of Silver (I). *Acc. Chem. Res.* **2000**, *33*, 831–840. [CrossRef] [PubMed]
128. Reppe, T.; Dressel, C.; Poppe, S.; Tschierske, C. Controlling spontaneous mirror symmetry breaking in cubic liquid crystalline phases by the cycloaliphatic ring size. *Chem. Commun.* **2020**, *56*, 711–714. [CrossRef] [PubMed]
129. Dressel, C.; Reppe, T.; Poppe, S.; Prehm, M.; Lu, H.; Zeng, X.; Ungar, G.; Tschierske, C. Helical networks of π-conjugated rods—A robust design concept for bicontinuous cubic liquid crystalline phases with achiral *Ia*3*d* and chiral *I*23 lattice. *Adv. Funct. Mater.* **2020**, submitted.
130. Reppe, T.; Poppe, S.; Cai, X.; Cao, Y.; Liu, F.; Tschierske, C. Spontaneous mirror symmetry breaking in benzil-based soft crystalline, cubic liquid crystalline and isotropic liquid phases. *Chem. Sci.* **2020**, *11*, 5902–5908. [CrossRef]
131. Wolska, J.M.; Wilk, J.; Pociecha, D.; Mieczkowski, J.; Gorecka, E. Optically Active Cubic Liquid Crystalline Phase Made of Achiral Polycatenar Stilbene Derivatives. *Chem. Eur. J.* **2017**, *23*, 6853–6857. [CrossRef]
132. Kutsumizu, S.; Yamada, Y.; Sugimoto, T.; Yamada, N.; Udagawa, T.; Miwa, Y. Systematic exploitation of thermotropic bicontinuous cubic phase families from 1,2-bis(aryloyl)hydrazine-based molecules. *Phys. Chem. Chem. Phys.* **2018**, *20*, 7953–7961. [CrossRef]
133. Kutsumizu, S.; Miisako, S.; Miwa, Y.; Kitagawa, M.; Yamamura, Y.; Saito, K. Mirror symmetry breaking by mixing of equimolar amounts of two gyroid phase-forming achiral molecules. *Phys. Chem. Chem. Phys.* **2016**, *18*, 17341–17344. [CrossRef]
134. Pescitelli, G.; Di Bari, L.; Berova, N. Application of electronic circular dichroism in the study of supramolecular systems. *Chem. Soc. Rev.* **2014**, *43*, 5211–5233. [CrossRef]
135. Gottarelli, G.; Lena, S.; Masiero, S.; Pieraccini, S.; Spada, G.P. The Use of Circular Dichroism Spectroscopy for Studying the Chiral Molecular Self-Assembly: An Overview. *Chirality* **2008**, *20*, 471–485. [CrossRef]
136. Jouvelet, B.; Isare, B.; Bouteiller, L.; von der Schoot, P. Direct probing of the free-energy penalty for helix reversal and chiral mismatches in chiral supramolecular polymers. *Langmuir* **2014**, *30*, 4570–4575. [CrossRef]
137. Lu, H.; Zeng, X.; Ungar, G.; Dressel, C.; Tschierske, C. The Solution of the Puzzle of Smectic-Q: The Phase Structure and the Origin of Spontaneous Chirality. *Angew. Chem. Int. Ed.* **2018**, *57*, 2835–2840. [CrossRef] [PubMed]
138. Brand, H.R.; Pleiner, H. Cubic and tetragonal liquid crystal phases composed of non-chiral molecules: Chirality and macroscopic properties. *Eur. Phys. J. E* **2019**, *42*, 142. [CrossRef] [PubMed]
139. Kajitani, T.; Kohmoto, S.; Yamamoto, M.; Kishikawa, K. Spontaneous Chiral Induction in a Cubic Phase. *Chem. Mater.* **2005**, *17*, 3812–3819. [CrossRef]
140. Quack, M. How important is parity violation for molecular and biomolecular chirality? *Angew. Chem. Int. Ed.* **2002**, *41*, 4618–4630. [CrossRef]
141. Chandrasekhar, S. Molecular homochirality and the parity-violating energy difference. A critique with new proposals. *Chirality* **2008**, *20*, 84–95. [CrossRef]
142. Ha, A.; Cohen, I.; Zhao, X.L.; Lee, M.; Kivelson, D. Supercooled Liquids and Polyamorphism. *J. Phys. Chem.* **1996**, *100*, 1–4. [CrossRef]
143. Anisimov, M.A.; Duška, M.; Caupin, F.; Amrhein, L.E.; Rosenbaum, A.; Sadus, R.J. Thermodynamics of Fluid Polyamorphism. *Phys. Rev. X* **2018**, *8*, 011004. [CrossRef]
144. Goodby, J.W.; Dunmur, D.A.; Collings, J.P. Lattice melting at the clearing point in frustrated Systems. *Liq. Cryst.* **1995**, *19*, 703–709. [CrossRef]

145. de las Heras, D.; Tavares, J.M.; Telo da Gama, M.M. Phase diagrams of binary mixtures of patchy colloids with distinct numbers of patches: The network fluid regime. *Soft Matter* **2011**, *7*, 5615–5626. [CrossRef]
146. Zhuang, Y.; Charbonneau, P. Recent Advances in the Theory and Simulation of Model Colloidal Microphase Formers. *J. Phys. Chem. B* **2016**, *120*, 7775–7782. [CrossRef]
147. Henrich, O.; Stratford, K.; Cates, M.E.; Marenduzzo, D. Structure of Blue Phase III of Cholesteric Liquid Crystals. *Phys. Rev. Lett.* **2011**, *106*, 107801. [CrossRef]
148. Brand, H.R.; Pleiner, H. On the influence of a network on optically isotropic fluid phases with tetrahedral/octupolar order. *Eur. Phys. J. E* **2017**, *40*, 34. [CrossRef] [PubMed]
149. Damer, B.; Deamer, D. The Hot Spring Hypothesis for an Origin of Life. *Astrobiology* **2020**, *20*, 429–452. [CrossRef] [PubMed]
150. Camprubí, E.; de Leeuw, J.W.; House, C.H.; Raulin, F.; Russell, M.J.; Spang, A.; Tirumalai, M.R.; Westall, F. The Emergence of Life. *Space Sci. Rev.* **2019**, *215*, 56. [CrossRef]
151. Toxvaerd, S. The role of the peptides at the origin of life. *J. Theor. Biol.* **2017**, *429*, 164–169. [CrossRef]
152. Gilbert, W. The RNA world. *Nature* **1986**, *319*, 618. [CrossRef]
153. Pressman, A.; Blanco, C.; Chen, I.A. The RNA World as a Model System to Study the Origin of Life. *Curr. Biol.* **2015**, *25*, R953–R963. [CrossRef]
154. Sutherland, J.D. The Origin of Life—Out of the Blue Angew. *Chem. Int. Ed.* **2016**, *55*, 104–121. [CrossRef]
155. Cafferty, B.J.; Fialho, D.M.; Hud., N.V. Searching for Possible Ancestors of RNA: The Self-Assembly Hypothesis for the Origin of Proto-RNA. In *Prebiotic Chemistry and Chemical Evolution of Nucleic Acids. Nucleic Acids and Molecular Biology*; Menor-Salván, C., Nicholson, A.W., Eds.; Springer-Nature: Cham, Switzerland, 2018; Volume 35.
156. Krishnamurthy, R. On the Emergence of RNA. *Isr. J. Chem.* **2015**, *55*, 837–850. [CrossRef]
157. Xu, J.; Tsanakopoulou, M.; Magnani, C.R.; Szabla, R.; Šponer, J.E.; Šponer, J.; Góra, R.W.; Sutherland, J.D. A prebiotically plausible synthesis of pyrimidine β-ribonucleosides and their phosphate derivatives involving photoanomerization. *Nat. Chem.* **2017**, *9*, 303–309. [CrossRef]
158. Nama, I.; Nama, H.G.; Zareb, R.N. Abiotic synthesis of purine and pyrimidine ribonucleosides in aqueous microdroplets. *Proc. Natl. Acad. Sci. USA* **2018**, *115*, 36–40. [CrossRef] [PubMed]
159. Becker, S.; Feldmann, J.; Wiedemann, S.; Okamura, H.; Schneider, C.; Iwan, K.; Crisp, A.; Rossa, M.; Amatov, T.; Carell, T. Unified prebiotically plausible synthesis of pyrimidine and purine RNA ribonucleotides. *Science* **2019**, *366*, 76–82. [CrossRef] [PubMed]
160. Hunter, C.A.; Sanders, J.K.M. The Nature of Interactions. *J. Am. Chem. Soc.* **1990**, *112*, 5525–5534. [CrossRef]
161. Martinez, C.R.; Iverson, B.L. Rethinking the term "pi-stacking". *Chem. Sci.* **2012**, *3*, 2191–2201. [CrossRef]
162. Krueger, A.T.; Kool, E.T. Model systems for understanding DNA base pairing. *Curr. Opin. Chem. Biol.* **2007**, *11*, 588–594. [CrossRef] [PubMed]
163. Joyce, G.F.; Schwartz, A.W.; Miller, S.L.; Orgel, L.E. The case for an ancestral genetic system involving simple analogues of the nucleotides. *Proc. Natl. Acad. Sci. USA* **1987**, *84*, 4398–4402. [CrossRef]
164. Chen, Y.; Ma, W. The origin of biological homochirality along with the origin of life. *PLoS Comput. Biol.* **2020**, *16*, e1007592. [CrossRef]
165. Sandars, P.G.H. A toy model for the generation of homochirality during polymerization. *Orig. Life Evol. Biosph.* **2003**, *33*, 575–587. [CrossRef]
166. Wattis, J.A.D.; Coveney, P.V. Symmetry-breaking in chiral polymerisation. *Orig. Life Evol. Biosph.* **2005**, *35*, 243–273. [CrossRef]
167. Jain, V.; Cheon, K.-S.; Tang, K.; Jha, S.; Green, M.M. Chiral Cooperativity in Helical Polymers. *Isr. J. Chem.* **2011**, *51*, 1067–1074. [CrossRef]
168. Baumgarten, J.L. Ferrochirality: A simple theoretical model of interacting, dynamically invertible, helical polymers, 2[†]. Molecular field approach: Supports and the details. *Macromol. Theory Simul.* **1995**, *4*, 1–43. [CrossRef]
169. Stich, M.; Hochberg, D. Mechanically Induced Homochirality in Nucleated Enantioselective Polymerization, Celia Blanco. *J. Phys. Chem. B* **2017**, *121*, 942–955. [CrossRef]

170. Buchs, J.; Vogel, L.; Janietz, D.; Prehm, M.; Tschierske, C. Chirality Synchronization of Hydrogen-Bonded Complexes of Achiral N-Heterocycles. *Angew. Chem. Int. Ed.* **2017**, *56*, 280–284. [CrossRef]
171. Karunakaran, S.C.; Cafferty, B.J.; Weigert-MuÇoz, A.; Schuster, G.B.; Hud, N.V. Spontaneous Symmetry Breaking in the Formation of Supramolecular Polymers: Implications for the Origin of Biological Homochirality. *Angew. Chem. Int. Ed.* **2019**, *58*, 1453–1457. [CrossRef]

Article

Chiral Oscillations and Spontaneous Mirror Symmetry Breaking in a Simple Polymerization Model

William Bock and Enrique Peacock-López *

Department of Chemistry, Williams College, Williamstown, MA 01267, USA; wb3@williams.edu
* Correspondence: epeacock@williams.edu

Received: 10 July 2020; Accepted: 19 August 2020; Published: 20 August 2020

Abstract: The origin of biological homochirality—defined as the preference of biological systems for only one enantiomer—has widespread implications in the study of chemical evolution and the origin of life. The activation—polymerization—epimerization—depolymerization (APED) model is a theoretical model originally proposed to describe chiral symmetry breaking in a simple dimerization system. It is known that the model produces chiral and chemical oscillations for certain system parameters, in particular, the preferential formation of heterochiral polymers. In order to investigate the effect of higher oligomers, our model adds trimers, tetramers, and pentamers. We report sustained oscillations of all chemical species and the enantiomeric excess for a wide range of parameter sets as well as the periodic chiral amplification of a small initial enantiomeric excess to a nearly homochiral state.

Keywords: chiral oscillations; spontaneous mirror symmetry breaking; origin of homochirality

---

1. Introduction

The origin of biological homochirality has attracted researchers' attentions for decades. At the center of this issue lies the question of how an optically inactive mixture—that is, a mixture containing equal amounts of two enantiomers—can react in such a way that propagates a small initial enantiomeric excess (*ee*) to form a product that is enantiopure. The quest for a mechanism that could have caused such spontaneous mirror symmetry breaking (SMSB) in a prebiotic environment has spurred decades of theoretical and experimental research. In Frank's seminal paper [1], he introduced a simple model capable of SMSB, in which enantiomers L and D catalyze their own formation, and they inhibit each other by forming the non-reactive heterochiral dimer LD. The latter step, which Frank termed "mutual antagonism", is crucial for chiral amplification, because the formation of the heterodimer is a greater penalty for the rarer enantiomer, allowing the common one to predominate.

Frank's paradigmatic scheme has inspired many variations and extensions, including several theoretical models investigating symmetry breaking in the famous Soai reaction [2]. In one such model, homochiral dimers are formed in addition to heterochiral ones and dimerization is reversible [3]. Models have also been proposed in which the monomers themselves are not autocatalytic, but instead homochiral dimers catalyze the formation of monomers of the same chirality [4,5].

The same paradigm has also been extended to systems involving longer polymers made up of amino acids or nucleotides. In one influential polymerization model, only the longest homochiral polymers catalyze the formation of chiral monomers, and mutual antagonism takes the form of enantiomeric cross inhibition, whereby the addition of a wrong-handed monomer to a growing chain halts elongation of that chain [6]. Many variations of this model have since been proposed [7,8], including one in which homochirality arises in concert with the emergence of life [9].

In this article, we study a toy polymerization model introduced by Plasson et al. that stands out from the aforementioned ones, in that it does not involve direct catalysis in monomer synthesis [10]. Instead, it introduces autocatalytic behavior through the presence of stereoselective epimerization reactions, whereby heterodimers LD and DL epimerize at the N-terminal residue to form homodimers DD and LL, respectively. In this case, the position adjacent to the epimerizing center can be considered an autocatalyst, because it converts the opposite enantiomer to its own chirality. These epimerization reactions, along with polymerization and hydrolysis, introduce both the autocatalytic behavior and mutual inhibition necessary for the destabilization of the racemic state [10,11].

The system is composed of activation, polymerization, epimerization, and depolymerization (APED) reactions between deactivated monomers L and D; activated monomers L* and D*; and dimers LL, DD, DL, and LD, where L and D represent enantiomers and DL and LD have the same chemical properties:

$$L \xrightarrow{a} L^* \qquad LL \xrightarrow{h} L+L \tag{1}$$
$$D \xrightarrow{a} D^* \qquad DD \xrightarrow{h} D+D \tag{2}$$
$$L^* \xrightarrow{b} L \qquad LD \xrightarrow{\beta h} L+D \tag{3}$$
$$D^* \xrightarrow{b} D \qquad DL \xrightarrow{\beta h} D+L \tag{4}$$

$$L+L^* \xrightarrow{p} LL \qquad LL \xrightarrow{\gamma e} DL \tag{5}$$
$$D+D^* \xrightarrow{p} DD \qquad DD \xrightarrow{\gamma e} LD \tag{6}$$
$$L+D^* \xrightarrow{\alpha p} DL \qquad DL \xrightarrow{e} LL \tag{7}$$
$$D+L^* \xrightarrow{\alpha p} LD \qquad LD \xrightarrow{e} DD \tag{8}$$

The model is contained within a closed system with a fixed concentration, and the rates of polymerization, depolymerization and epimerization are different for their homo- and heterochiral counterparts (i.e., they are stereoselective).

Although Plasson et al.'s report searches for the monotonic emergence of chiral states, they note that oscillations of all chemical species, and the enantiomeric excess, occur when heterochiral dimers are strongly preferred. These oscillations are the focus of Stich et al.'s report, which performs a comprehensive bifurcation analysis and finds that oscillations occur for a wide range of parameter sets [12]. Such chiral oscillatory phenomena have important implications for the transmission of chirality in chemical systems: in contrast to models for SMSB that involve the monotonic emergence of chiral steady states, systems involving chiral oscillations add another layer of stochasticity to the final sign of chirality that is ultimately transmitted to the system. This is because the memory of the sign of any initial fluctuation is erased by subsequent oscillations. In recent years, chiral oscillations have been of growing interest in theoretical models [12,13], especially in light of experimental evidence of chiral oscillations in the polymerization of some amino acids [14].

The APED model is a powerful tool for studying chiral oscillatory phenomena because its simplicity makes it amenable to the comprehensive mathematical analysis of its parameters, which can provide insight into the driving forces behind real chemical systems involving chiral oscillations. Despite laying the groundwork for describing chiral oscillations in polymerization systems, its restriction to dimers makes it unable to encapsulate the nuances that result from the formation of higher oligomers. One experimental study [15], for example, shows that the stereoselectivities of epimerization and hydrolysis for some amino acids differ between di- and tripeptides. Introducing trimers to the model takes one step closer to capturing such nuances while preserving the simplicity and applicability of the original APED scheme.

This article is organized as follows. In Section 2, we introduce the expanded APED model and demonstrate chiral and chemical oscillations in the expanded model. In Section 3, we analyze the bifurcation diagrams of several system parameters and then expand the model further to examine the onset of oscillations in the presence of tetramers and pentamers. In Section 4, we explore the minimum initial enantiomeric excess required to produce oscillations in the minimal and expanded APED models. We also demonstrate large amplitude oscillations in the overall enantiomeric excess, which can periodically achieve near homochirality ($ee > 0.98$). Finally, we close the article with a discussion of the results in Section 5.

2. The Expanded APED Model

The APED model was originally introduced to describe the monotonic chiral symmetry breaking of peptides, in contrast to the chiral oscillations presented in this report (and originally presented by Stich et al.). This discrepancy results from Plasson et al.'s assumption that amino acids tend to condense into homochiral peptides faster than heterochiral ones. Although this can be true [16,17], experimental evidence involving the condensation of phenylglycine reports damped chiral and chemical oscillations by peptides, and it has been suggested that the oscillations are caused by an APED-like system that favors the formation of heterochiral peptides [14].

To our knowledge, the numerous applications and variations of the APED model have all been limited to dimers [10,12,14]. Although the APED is strictly a toy model [18], the inclusion of higher oligomers can more accurately simulate real polymerization systems while preserving the simplicity and generality of the APED scheme, which is what lends itself well to the systematic analysis of the parameters (presented in Section 3).

We assume that polymerization in the expanded model is restricted by enantiomeric cross-inhibition; that is, only homochiral chains are capable of polymerization, and the addition of a monomer of the opposite chirality halts elongation of that chain. This phenomenon has been shown in the oligomerization of activated mononucleotides [19]. Furthermore, in order to maintain the model's simplicity, polymerization is unidirectional and is limited to monomer addition. The reactions we add to the system are represented by the following equations.

$$LL + L^* \xrightarrow{p_2} LLL \qquad DLL \xrightarrow{\beta_2 h_2} D + LL \tag{9}$$

$$DD + D^* \xrightarrow{p_2} DDD \qquad LDD \xrightarrow{\beta_2 h_2} L + DD \tag{10}$$

$$LL + D^* \xrightarrow{\alpha_2 p_2} DLL \qquad DLL \xrightarrow{e_2} LLL \tag{11}$$

$$DD + L^* \xrightarrow{\alpha_2 p_2} LDD \qquad LDD \xrightarrow{e_2} DDD \tag{12}$$

$$LLL \xrightarrow{h_2} L + LL \qquad LLL \xrightarrow{\gamma_2 e_2} DLL \tag{13}$$

$$DDD \xrightarrow{h_2} D + DD \qquad DDD \xrightarrow{\gamma_2 e_2} LDD \tag{14}$$

Our expansion introduces homo- and heterochiral trimers, and the new reactions include homochiral polymerization (rate p_2), heterochiral polymerization (rate $\alpha_2 p_2$), homochiral hydrolysis (rate h_2), heterochiral hydrolysis (rate $\beta_2 h_2$), homochiral epimerization (rate e_2), and heterochiral epimerization (rate $\gamma_2 e_2$). Similar to the original APED model, mass is conserved so that the total concentration c = [L] + [D] + [L*] + [D*] + 2([LL] + [DD] + [LD] + [DL]) + 3([LLL] + [DDD] + [LDD] + [DLL]) is constant. The enantiomeric excess is defined as ee = [L] + [L*] + 2[LL] + 3[LLL] + [DLL] - ([D] + [D*] + 2[DD] + 3[DDD] + [LDD])/c.

The trimer reactions in conjunction with the original APED model transcribe into the following set of ordinary differential equations,

$$\frac{d[L]}{dt} = -a[L] - p_1[L][L^*] - \alpha_1 p_1[L][D^*] + 2h_1[LL] + b[L^*] + \beta_1 h_1[DL] + \beta_1 h_1[LD] + \beta_2 h_2[LDD] + h_2[LLL] \quad (15)$$

$$\frac{d[D]}{dt} = -a[D] - p_1[D][D^*] - \alpha_1 p_1[D][L^*] + 2h_1[DD] + b[D^*] + \beta_1 h_1[DL] + \beta_1 h_1[LD] + \beta_2 h_2[DLL] + h_2[DDD] \quad (16)$$

$$\frac{d[L^*]}{dt} = a[L] - b[L^*] - \alpha_1 p_1[L^*][D] - p_2[LL][L^*] - p_1[L][L^*] - \alpha_2 p_2[DD][L^*] \quad (17)$$

$$\frac{d[D^*]}{dt} = a[D] - b[D^*] - \alpha_1 p_1[D^*][L] - p_2[DD][D^*] - p_1[D][D^*] - \alpha_2 p_2[LL][D^*] \quad (18)$$

for the dimers,

$$\frac{d[LL]}{dt} = p_1[L][L^*] - h_1[LL] + e_1[DL] - p_2[LL][L^*] - \gamma_1 e_1[LL] + h_2[LLL] - \alpha_2 p_2[LL][D^*] + \beta_2 h_2[DLL] \quad (19)$$

$$\frac{[DD]}{dt} = p_1[D][D^*] - h_1[DD] + e_1[LD] - p_2[DD][D^*] - \gamma_1 e_1[DD] + h_2[DDD] - \alpha_2 p_2[DD][L^*] + \beta_2 h_2[LDD] \quad (20)$$

$$\frac{[DL]}{dt} = \alpha_1 p_1[L][D^*] - e_1[DL] - \beta_1 h_1[DL] + \gamma_1 e_1[LL] \quad (21)$$

$$\frac{[LD]}{dt} = \alpha_1 p_1[D][L^*] - e_1[LD] - \beta_1 h_1[LD] + \gamma_1 e_1[DD] \quad (22)$$

and finally for the trimers,

$$\frac{[LLL]}{dt} = p_2[LL][L^*] - h_2[LLL] + e_2[LLD] - \gamma_2 e_2[LLL] \quad (23)$$

$$\frac{[DDD]}{dt} = p_2[DD][D^*] - h_2[DDD] + e_2[DDL] - \gamma_2 e_2[DDD] \quad (24)$$

$$\frac{[DLL]}{dt} = \alpha_2 p_2[LL][D^*] - \beta_2 h_2[DLL] - e_2[DLL] + \gamma_2 e_2[LLL] \quad (25)$$

$$\frac{[LDD]}{dt} = \alpha_2 p_2[DD][L^*] - \beta_2 h_2[LDD] - e_2[LDD] + \gamma_2 e_2[DDD] \quad (26)$$

Oscillations for the expanded APED model are illustrated in Figure 1, which displays the concentration of the homochiral trimers DDD and LLL as well as the overall enantiomeric excess. For parameter values, we choose $\alpha_1 = \alpha_2 = 50$, $a = p_1 = h_1 = e_1 = p_2 = h_2 = e_2 = 1$, $b = \beta_1 = \gamma_1 = \beta_2 = \gamma_2 = 0$, $c = 0.5$, and we keep a = 1 and b = 0 for the remainder of the article. In this case, we use an initial enantiomeric excess of $ee_{init} = 0.01$, and we see that this initial excess is propagated to oscillations ranging from $ee \approx -0.6$ to 0.6 within 200 dimensionless time increments. In Section 4, we show that the amplitude of oscillations in ee increases substantially as higher oligomers are introduced to the system.

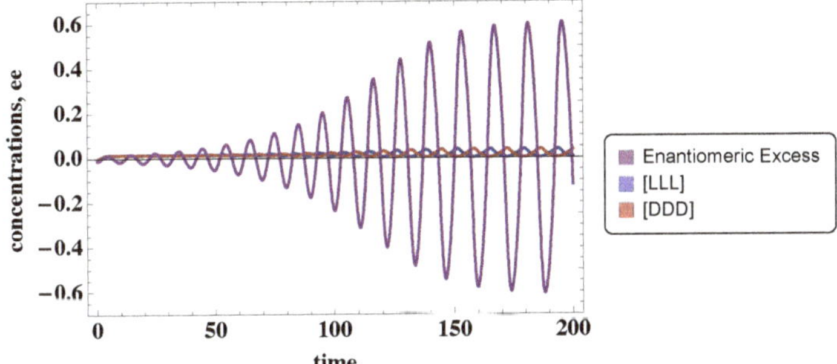

Figure 1. Stable oscillations of the concentrations of LLL and DDD as well as the enantiomeric excess (ee) for the parameter values $\alpha_1 = \alpha_2 = 20, a = h_1 = e_1 = p_1 = p_2 = h_2 = e_2 = 1, b = \beta_1 = \gamma_1 = \beta_2 = \gamma_2 = 0$, $c = 0.5$ and $ee_{init} = 0.01$. For simplicity we only display [LLL] and [DDD], although all concentrations oscillate. The units for time are dimensionless, because we do not use literature values for the reaction rates.

3. Bifurcation Analysis

3.1. Trimer Model

Using Stich et al.'s bifurcation analysis of the minimal APED model as a platform, in this section we present our bifurcation analysis of several system parameters to determine how general is the appearance of oscillations in the expanded model. In particular, we focus on the relative stereoselectivities of polymerization, epimerization, and hydrolysis, namely, α_1 and α_2, γ_1 and γ_2, and β_1 and β_2. In addition to the bifurcation diagrams for each individual parameter, we also present joint bifurcation diagrams where the stereoselectivities are held constant for dimers and trimers (i.e., $\alpha_1 = \alpha_2$, $\gamma_1 = \gamma_2$, $\beta_1 = \beta_2$). These joint-parameter diagrams are useful for exploring the generality of oscillations, although it is worth noting that the stereoselectivities of peptide polymerization, epimerization, and hydrolysis are not necessarily independent of chain length [15].

In our bifurcation diagrams, we track the onset of oscillations through the concentration of the homochiral trimer DDD. In Figure 2a, for example, the upper line of points represents the maximum concentration of DDD at each value of α_1—that is, the peak of the chemical oscillations—and the bottom represents the minimum concentration of DDD at each value of α_1. The region of the graph where there is only one line (i.e., $[\text{DDD}]_{max}$=$[\text{DDD}]_{min}$), there are no oscillations. All bifurcation diagrams were created using an initial ee of 0.01.

Similar to the original APED model, oscillations in the trimer model depend on the preferential polymerization of heterochiral peptides. With the addition of trimers, though, the minimum values for heterochiral polymerization (α_1 and α_2) are significantly lower than that for the dimer model. The bifurcation diagram for α_1 indicates that stable oscillations begin at around α_1 = 5 for α_2 = 20, which is more than 10 (dimensionless) units below the minimum value for α_1 in the original APED model, which is just over 16. Similarly, Figure 2b depicts that the minimum value for α_2 when α_1 is held constant at 20 is just under 6, also well below the original APED's 16. Finally, in order to consider the case for which the rates of homo- and heterochiral polymerization are the same, the bifurcation diagram in Figure 2c indicates that the joint parameter bifurcates at 10, also significantly lower than in the original APED. Increasing the allowed range of values for heterochiral polymerization allows the model to apply to a wider range of real peptide-forming systems.

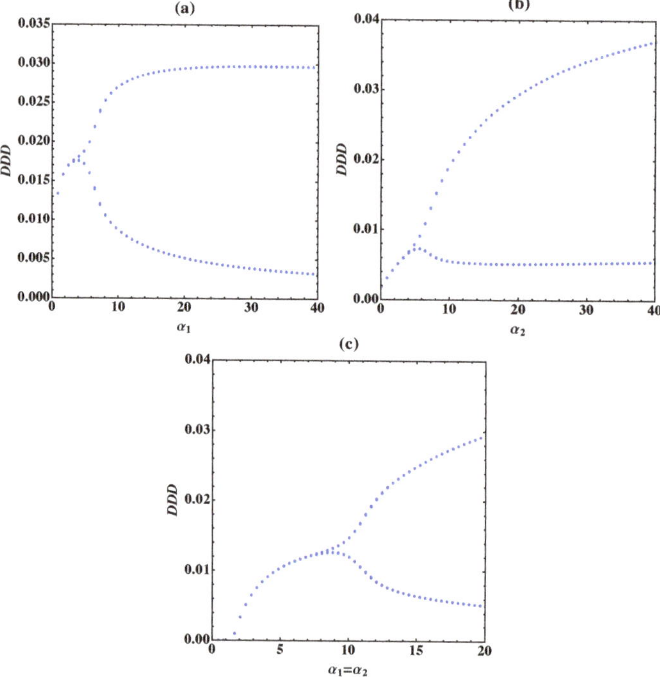

Figure 2. The bifurcation diagrams for α_1 (**a**) and α_2 (**b**) and the joint parameter where $\alpha_1 = \alpha_2$ (**c**) illustrate that heterodimers and heterotrimers must be formed preferentially for oscillations to occur. Other parameters are $a = p_1 = h_1 = e_1 = p_2 = h_2 = e_2 = 1$, $b = \beta_1 = \gamma_1 = \beta_2 = \gamma_2 = 0$, $\alpha_2 = 20$ (**a**), $\alpha_1 = 20$ (**b**), $c = 0.5$, $ee_{init} = 0.01$.

To check the impact of the stereoselectivity of epimerization, in Figure 3 we explore the bifurcation diagrams of γ_1, γ_2, and the joint parameter where they share the same value. Figure 3a illustrates that the region for stable oscillations closes around $\gamma_1 \approx 1.15$. The fact that the region extends past $\gamma_1 = 1$ indicates that stable oscillations occur even in the case of unity, which is the absence of any stereoselectivity of dimer epimerization. Similarly, in Figure 3b,c, which display the bifurcation diagrams for γ_2 and the joint parameter between γ_1 and γ_2, the regions of stable oscillations close at around $\gamma_2 \approx 1.5$ and $\gamma_1 = \gamma_2 \approx 1.15$, respectively. These diagrams suggest that epimerization need not be stereoselective for dimers nor trimers to produce oscillations.

The bifurcation diagrams in Figure 4 indicate that the allowed range of values for β_1, β_2, and the joint parameter where $\beta_1 = \beta_2$ close at 0.27, 0.95, and 0.23, respectively. These results suggest that the allowed range of values are more restricted for the stereoselectivities of hydrolysis. Despite this restrictiveness, Figure 4d displays another version of the joint bifurcation diagram for β_1 and β_2 but with the parameter values $h_1 = h_2 = 0.2$, $e_1 = e_2 = 0.1$. We note that a peptide system for which the rate of polymerization is faster than hydrolysis and hydrolysis is faster than epimerization is chemically realistic [15]. With this in mind, Figure 4d demonstrates that oscillations also occur in the absence of stereoselective hydrolysis for dimers and trimers.

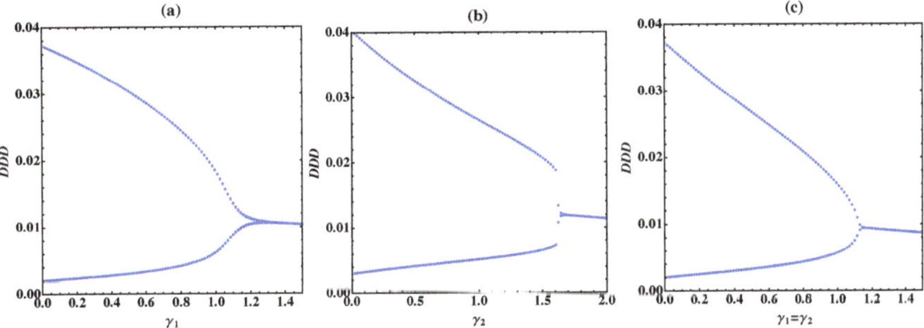

Figure 3. The bifurcation diagrams for γ_1 (**a**), γ_2 (**b**), and the joint parameter where γ_1=γ_2. Panel (**c**) illustrates the flexibility of the range of values for the stereoselectivity of epimerization that give rise to oscillations. Under these parameter values the range of allowed values for stereoselectivity of epimerization extend past unity. Other parameters are $a = p_1 = h_1 = p_2 = h_2 = e_2 = 1$, $b = \beta_1 = \beta_2 = \gamma_1 = \gamma_2 = 0, \alpha_1 = \alpha_2 = 50, e_1 = 0.5$ (**a**) and (**c**); $e_1 = 1$ (**b**), c = 0.5, $ee_{init} = 0.01$.

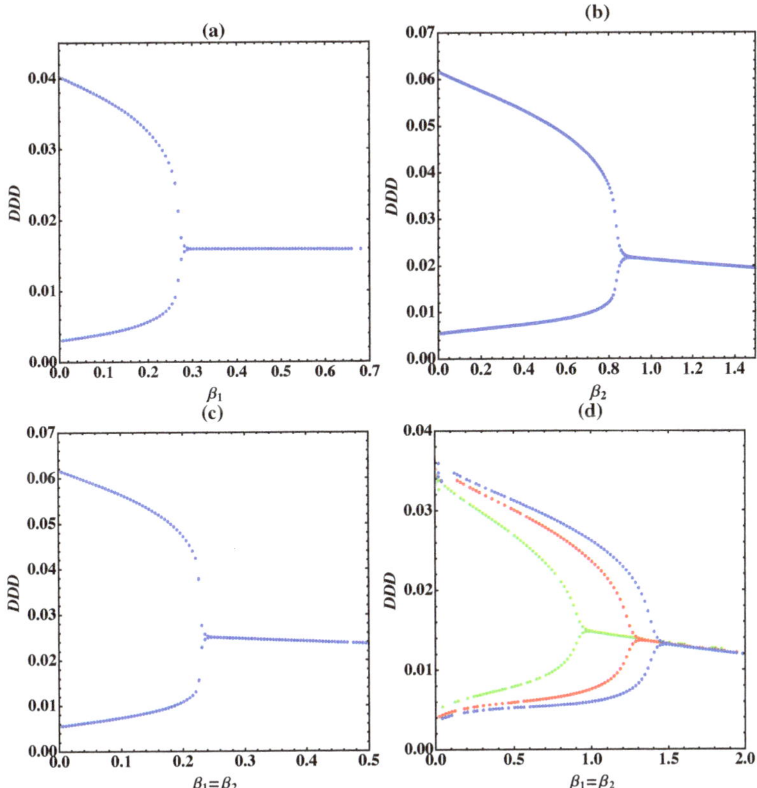

Figure 4. The bifurcation diagrams of β_1 (**a**), β_2 (**b**), and the joint parameter where β_1=β_2 (**c**) depicts the range of parameter values for the stereoselectivity of hydrolysis that result in oscillations. Under this set of parameter values, the range of allowed values is not as flexible for hydrolysis as for epimerization. Other parameters are $a = p_1 = h_1 = e_1 = p_2 = e_2 = 1$, $b = \gamma_1 = \gamma_2 = \beta_2 = 0$, $\alpha_1 = \alpha_2 = 50$, $h_2 = 1$ (**a**), $h_2 = 0.5$ (**b**), and (**c**), $c = 0.5$, $ee_{init} = 0.01$. Panel (**d**) is the joint bifurcation diagram of β_1 and β_2 with the parameters $h_1 = h_2 = 0.2, e_1 = e_2 = 0.1$. The three values are $\alpha_1 = \alpha_2 = 50$ (green), $\alpha_1 = \alpha_2 = 100$ (red), $\alpha_1 = \alpha_2 = 150$ (blue).

To conclude our bifurcation analysis for the trimer model, we explore the relationship between the stereoselectivities of hydrolysis and epimerization using an analogous parameter set to Figure 4d in order to find stable oscillations in the absence of any stereoselective epimerization and hydrolysis. Figure 5 shows the joint bifurcation diagrams of β_1 and γ_1 as well as β_2 and γ_2, that is, the case where the stereoselectivities of hydrolysis and epimerization are the same for dimers and trimers, respectively. Although these two stereoselectivities are not directly related chemically, the bifurcation diagrams illustrate that oscillations still occur if both heterochiral hydrolysis and epimerization are preferred for di- or tripeptides. Finally, in order to demonstrate oscillations in the absence of stereoselectivity for all hydrolysis and epimerization terms, Figure 5c revisits the joint bifurcation diagram for γ_2 and β_2, but with a parameter set for which $\gamma_1 = \beta_1 = 1$. For the other parameters, we choose $h_1 = h_2 = 0.1$ and $e_1 = e_2 = 0.02$ and we consider three values for α_1 and α_2. This specific example demonstrates that homochiral epimerization and depolymerization are not necessary conditions for sustained oscillations.

3.2. Tetramer and Pentamer Models

The bifurcation analysis in Section 3.1 shows that oscillations occur in the trimer APED model for a wide range of parameters. Here, we expand the model further to include homo- and heterochiral tetramers and pentamers in order to investigate the onset of oscillations in the presence of higher oligomers. The reactions we add include homochiral polymerization (rates p_3, p_4), heterochiral polymerization (rates $\alpha_3 p_3, \alpha_4 p_4$), homochiral hydrolysis (rates h_3, h_4), heterochiral hydrolysis (rates $\beta_3 h_3$, $\beta_4 h_4$), homochiral epimerization (rates e_3, e_4), and heterochiral epimerization (rates $\gamma_3 e_3$, $\gamma_4 e_4$). Instead of analyzing each individual parameter (as in Section 3.1), we focus our discussion on the joint parameters for polymerization, epimerization, and hydrolysis.

We find that the tetramer and pentamer models share many of the same characteristics as the minimal and trimer models. The bifurcation diagram in Figure 6a indicates that preferential formation of heterochiral oligomers is still necessary for oscillations, although the range of allowed values for the joint parameter for heterochiral polymerization closes at 7.75 for the tetramer model ($\alpha_1 = \alpha_2 = \alpha_3$) and 7.25 for the pentamer model ($\alpha_1 = \alpha_2 = \alpha_3 = \alpha_4$), each of which is less than the minimum values for the trimer model ($\alpha_1 = \alpha_2 = 10$) and minimal model ($\alpha_1 = 16$). This trend suggests that the requirement for strongly preferential heterochiral polymerization becomes less restrictive as oligomers of increasing length are incorporated into the system.

The bifurcation diagrams in Figure 6b,c indicate that, similar to the minimal and trimer models, the tetramer and pentamer models also favor homochiral epimerization and hydrolysis. Nevertheless, Figure 6b illustrates that the range of allowed values for the stereoselectivity of epimerization closes at 1.38 and 1.21 for the tetramer and pentamer models, respectively. The fact that both values extend past unity means that epimerization in both systems need not be stereoselective for oscillations to occur. Similarly, Figure 6c, which uses the analogous parameter set as in Figure 4d, demonstrates that oscillations occur in the absence of stereoselective hydrolysis for both the tetramer and pentamer models.

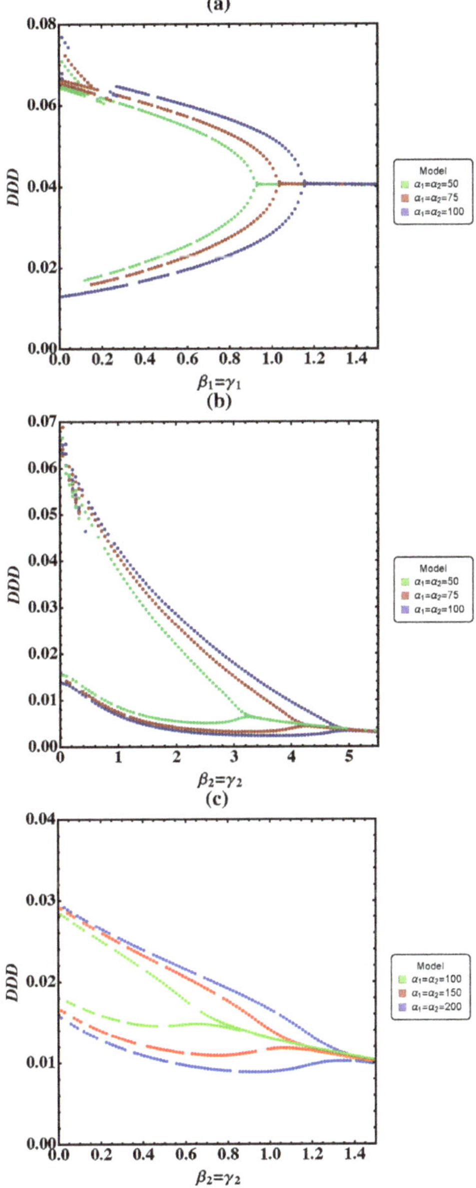

Figure 5. The joint bifurcation diagrams for γ_1 and β_1 (**a**) and γ_2 and β_2 (**b**) illustrate that sustained oscillations occur even when dimers or trimers favor homochiral epimerization and hydrolysis (i.e., $\gamma_1 = \beta_1 > 1$, $\gamma_2 = \beta_2 > 1$). Other parameters are $a = p_1 = p_2 = c = 1$, $h_1 = h_2 = 0.2, e_1 = e_2 = 0.1$, $\gamma_2 = \beta_2 = 0$ (**a**), $\gamma_1 = \beta_1 = 0$ (**b**), and $ee_{init} = 0.01$. Panel (**c**) is the joint bifurcation diagram for γ_2 and β_2 with a different set of parameters, including $\gamma_1 = \beta_1 = 1$, $h_1 = h_2 = 0.1$ $e_1 = e_2 = 0.02$. This diagram shows that oscillations occur even in the absence of stereoselectivity for all hydrolysis and epimerization terms.

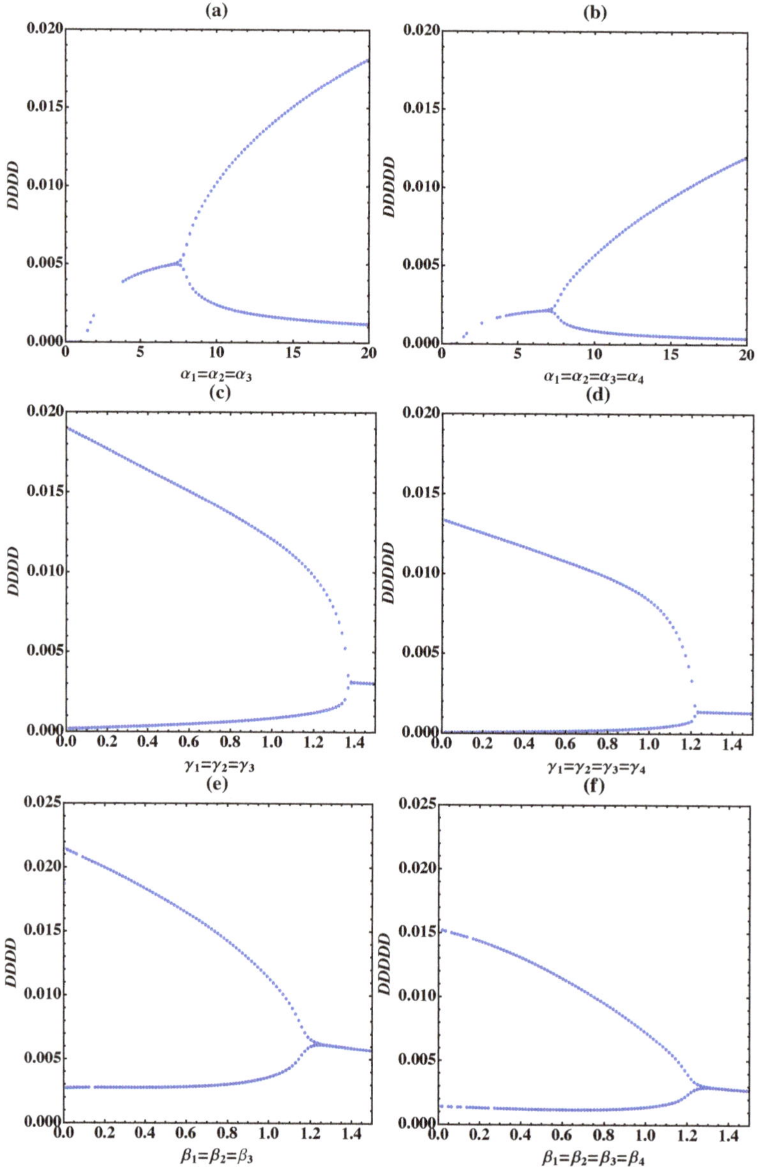

Figure 6. The joint bifurcation diagrams for the tetramer (left) and pentamer (right) activation—polymerization—epimerization—depolymerization (APED) models for polymerization (panels (**a**,**b**)), epimerization (panels (**c**,**d**)), and hydrolysis (panels (**e**,**f**)). Parameters in (**a**,**b**) are $a = p_{1,2,3,4} = h_{1,2,3,4} = e_{1,2,3,4} = 1$, $b = \beta_{1,2,3,4} = \gamma_{1,2,3,4} = 0$, $ee_{init} = 0.01$. Parameters in (**c**,**d**) are same as (**a**,**b**) except for $\alpha_{1,2,3,4} = 50$ and $e_{1,2,3,4} = 0.5$, and parameters in (**e**,**f**) are the same as (**c**,**d**) except for $e_{1,2,3,4} = 0.1$ and $h_{1,2,3,4} = 0.2$.

4. Minimum Initial Enantiomeric Excess

Here, we explore the minimum initial enantiomeric excess (*ee*) required for oscillations in the minimal and expanded APED models. For the APED model and its expansions studied in this report, each unique oscillation-producing parameter set results in oscillations of a certain amplitude, and above a critical value, the amplitude of the oscillations is independent of the initial excess. If, on the other hand, the initial excess is below the critical value, then the system reaches a dead state where none of the concentrations oscillate. Analysis of the minimum initial *ee* is relevant when considering a chemical system's capacity to undergo spontaneous mirror symmetry breaking, because systems with smaller minimums will respond to more subtle chiral perturbations, such as statistical chiral fluctuations or external chiral influences.

Throughout our analysis, in order to standardize the conditions for each system, we set the rates and stereoselectivities of polymerization, epimerization, and hydrolysis equal across all models. We also hold the concentrations of each system to $c = 1$. We first look at the cases for which $\alpha_1 = \alpha_2 = \alpha_3 = \alpha_4 = 25$, 50, 75, 100, 125, and 150, and the graph of our results is presented in Figure 7a. We find that the minimum initial *ee* depends on chain length for all values of $\alpha_1 = \alpha_2 = \alpha_3 = \alpha_4$, and the critical value generally decreases as chain length increases. The exception to this trend occurs at $\alpha_1 = \alpha_2 = \alpha_3 = \alpha_4 = 125$ where the minimum value of the tetramer model falls to $20^{-10.8}$ compared to 20^{-9} in the pentamer model. Of the parameter sets studied, the minimum initial *ee* for the pentamer model is on average 2.5 orders of magnitude lower than that of the minimal model.

In Figure 7b, we check the impact of the stereoselectivity of epimerization on the minimum initial *ee* and find a similar trend to Figure 7a. Namely, the critical value generally decreases as chain length increases, especially for low values of $\gamma_1 = \gamma_2 = \gamma_3 = \gamma_4$ (i.e., when homochiral epimerization is more strongly favored). The exceptions to this trend occur at $\gamma_1 = \gamma_2 = \gamma_3 = \gamma_4 = 0.3$ where the minimum initial excess for the trimer model drops below that of the tetramer model, and at $\gamma_1 = \gamma_2 = \gamma_3 = \gamma_4 = 0.5$ where the critical values for both the trimer and tetramer models fall below that of the pentamer model.

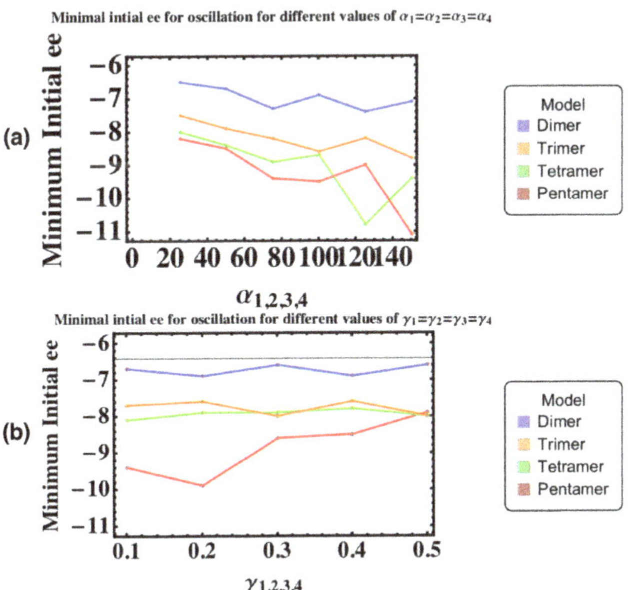

Figure 7. Plots of the minimum initial enantiomeric excesses (ee) required for oscillations and amplification in the minimal and expanded APED models for varying $\alpha_1 = \alpha_2 = \alpha_3 = \alpha_4$ (**a**) and varying $\gamma_1 = \gamma_2 = \gamma_3 = \gamma_4$ (**b**). Other parameters are $a = p_{1,2,3,4} = h_{1,2,3,4} = c = 1$, $e_{1,2,3,4} = 0.5$, $b = \beta_{1,2,3,4} = \gamma_{1,2,3,4} = 0$, and all other stereoselectivities (represented by greek letters) are zero. Parameters for (**b**) are same as (**a**) with $\alpha_{1,2,3,4} = 50$.

This preliminary analysis demonstrates that the initial *ee* required for oscillations and chiral amplification in the APED model depends both on the specific parameter set and the length of the largest oligomer allowed in the system.

Finally, to complete our study, we explore the amplitudes and periods of oscillations in the overall enantiomeric excess for the minimal and expanded APED models (Figure 8). For parameters we choose $\alpha_1 = \alpha_2 = \alpha_3 = \alpha_4 = 50$, $e_1 = e_2 = e_3 = e_4 = 0.5$, $c = 0.35$ and $ee_{init} = 10^{-6}$, and we see that the amplitude of the chiral oscillations increases as higher oligomers are introduced to the system. Specifically, oscillations in the dimer model range from -0.62 to 0.62, compared to -0.98 to 0.98 in the pentamer model. Furthermore, we also observe an inverse relationship between oligomer length and frequency, and the period of oscillations in the pentamer model is more than three times that of the dimer one.

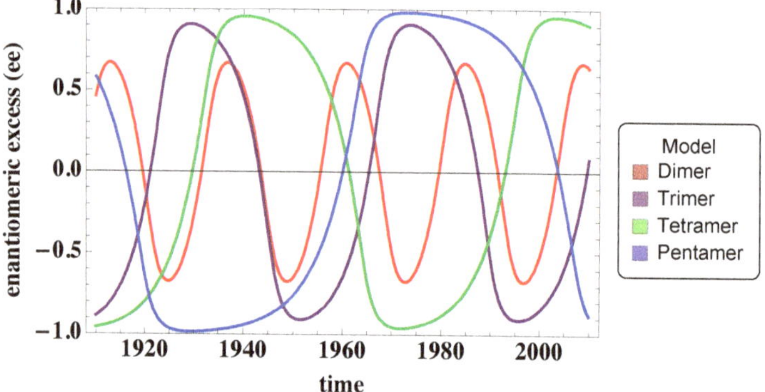

Figure 8. Oscillations in overall enantiomeric excess for the minimal and expanded APED models. Each simulation began with an initial *ee* of 10^{-6}, and the amplitudes of the oscillations for each model are as follows; dimer: -0.62, 0.62; trimer: -0.91, 0.91; tetramer: -0.96, 0.96; pentamer: -0.98, 0.98. Other parameters are $a = p_{1,2,3,4} = h_{1,2,3,4} = 1$, $e_{1,2,3,4} = 0.5$, $b = \beta_{1,2,3,4} = \gamma_{1,2,3,4} = 0$, $\alpha_{1,2,3,4} = 50$, $c = 0.35$. Similarly to Figure 1, the units of time are dimensionless.

5. Conclusions

In this paper, we have expanded the minimal APED model to include trimers, tetramers, and pentamers and have studied the range of parameter sets leading to chiral oscillations. In particular, we have analyzed the bifurcation diagrams of several system parameters including the stereoselectivities of polymerization, epimerization, and hydrolysis. We have also begun an investigation of the minimum initial enantiomeric excess required to produce oscillations in the minimal and expanded models.

Principally, we find that oscillations in the expanded APED models have similar properties to those in the original one: namely, oscillations are favored for heterochiral polymerization and homochiral depolymerization and epimerization. Additionally, we show that stable oscillations occur for a wide range of parameters. In particular, β_1, β_2, β_3, and β_4 as well as γ_1, γ_2, γ_3, and γ_4 can also be chosen at unity, indicating that hydrolysis and epimerization need not be stereoselective for oscillations to occur. On the other hand, heteropolymers still need to be formed preferentially, although the allowed range of values for α_1, α_2, α_3, and α_4 extends closer to unity with the inclusion of higher oligomers.

In addition to analyzing the effects of the stereoselectivities of polymerization, epimerization and hydrolysis on oscillations, using the trimer model we also explore the onset of oscillations as the stereoselectivities of hydrolysis and epimerization both approach unity. Although preferential homochiral epimerization and hydrolysis are favorable for oscillations, we show that oscillations occur even when all hydrolysis and epimerization terms are chosen at unity. While these rates are not directly related, this observation is noteworthy in light of experimental evidence suggesting that amino acid chains favor heterochiral epimerization and hydrolysis [15]. Moreover, we find that the range of allowed values for the stereoselectivities of epimerization and hydrolysis expands when those values vary with chain length. This is a significant observation when taking into consideration Danger et al.'s observation that homochiral chains are stabilized as chain length increases [15], meaning that in real systems of amino acids, even if γ_1 and β_1 must be greater than unity, $\gamma_{2,3,4}$ and $\beta_{2,3,4}$ are likely closer to unity, if not below.

Aside from the bifurcation analysis, another important property of any system involving SMSB—oscillatory or otherwise—is the minimum initial enantiomeric excess required to activate its propagation. We show that this critical value depends on the parameter set and the chain length of the longest polymer in the system. We also demonstrate that the critical value can be significantly lower when higher oligomers are incorporated into the system, although further investigation will be needed to determine the full effect of chain length.

Additionally, we show that chain length affects the amplitude and frequency of oscillations in the overall *ee*, and that the APED system produces more robust and longer lasting oscillations as higher oligomers are introduced. This discrepancy in amplitude is caused in large part by the fact that the hetero-oligomers in the dimer model are achiral, while those in the expanded models are not. This is because for the heterodimers LD and DL, the two enantiomers cancel each other out. In the expanded models, however, even though the two terminal monomers on the heterochiral chains still cancel each other out, the remaining ones contribute to the overall chirality in the system. As a result, each hetero-trimer contributes one molecule of chirality, the hetero-tetramers contribute two and the hetero-pentamers contribute three. Accordingly, the pentamer model experiences the largest amplitude oscillations, and it can temporarily, and periodically, achieve near homochirality from a small initial enantiomeric excess of 10^{-6}. Such large amplitude chiral oscillations have been studied previously in a polymerization model closed to matter and energy flow, in which the length of the homochiral chain formed determines the amplitude of the chiral oscillations [13]. They have also been demonstrated in a theoretical model designed to model symmetry breaking in the Soai reaction [20]. The amplitude, period, and minimum initial *ee* for oscillations are all important features when considering a chemical system's ability to amplify small chiral perturbations—whether they be caused by the presence of chiral crystals such as sodium chlorate [21], circularly polarized light [22], or another external chiral influence—in a prebiotic environment.

Author Contributions: Conceptualization, E.P.-L. and W.B.; methodology, E.P.-L.; software, W.B. and E.P.-L.; validation, W.B., E.P.-L.; formal analysis, W.B.; investigation, W.B.; resources, E.P.-L.; writing—original draft preparation, W.B.; writing—review and editing, W.B. and E.P.-L.; supervision, E.P.-L.; project administration, E.P.-L.; funding acquisition, E.P.-L. All authors have read and agreed to the published version of the manuscript.

Funding: This research received no external funding.

Acknowledgments: The authors would like to acknowledge the support from the summer research program at Williams College.

Conflicts of Interest: The authors declare no conflict of interest.

References

1. Frank, F.C. On spontaneous asymmetric synthesis. *Biochim. Biophys. Acta* **1953**, *11*, 459–463. [CrossRef]
2. Soai, K.; Shibata, T.; Morioka, H.; Choji, K. Asymmetric autocatalysis and amplification of enantiomeric of a chiral molecule. *Nature* **1995**, *378*, 767–768. [CrossRef]
3. Ribo, J.M.; Hochberg, D. Stability of racemic and chiral steady states in open and closed chemical systems. *Phys. Lett. A* **2008**, *373*, 111–122. [CrossRef]
4. Blackmond, D.G. Asymmetric amplification and its implications for the origin of homochirality. *Proc. Natl. Acad. Sci. USA* **2004**, *101*, 5732–5736. [CrossRef]
5. Islas, J.R.; Lavabre, D.; Grevy, J.M.; Lamoneda, R.H.; Cabrera, H.R.; Micheau, J.C.; Buhse, T. Mirror-symmetry breaking in the Soai reaction: A kinetic understanding. *Proc. Natl. Acad. Sci. USA* **2005**, *102*, 13743–13748. [CrossRef]

6. Sandars, P.G.H. A toy model for the generation of homochirality during polymerization. *Orig. Life Evol. Biosph.* **2003**, *33*, 575–587. [CrossRef]
7. Brandenburg, A.; Anderson, A.C.; Höfner, S.; Nilsson, M. Homochiral growth through enantiomeric cross inhibition. *Orig. Life. Evol. Biosph.* **2005**, *35*, 225–241. [CrossRef]
8. Brandenburg, A.; Anderson, A.C.; Höfner, S.; Nilsson, M. Dissociation in a polymerization model of homochirality. *Orig. Life. Evol. Biosph.* **2005**, *35*, 507–521. [CrossRef]
9. Wu, M.; Walker, S.I.; Higgs, P.G. Autocatalytic replication and homochirality in biopolymers: Is homochirality a requirement of life or a result of it? *Astro* **2012**, *12*, 818–829. [CrossRef]
10. Plasson, R.; Bersisni, H.; Commeyras, A. Recycling Frank: Spontaneous emergence of homochirality in noncatalytic systems. *Proc. Natl. Acad. Sci. USA* **2004**, *48*, 16733–16738. [CrossRef]
11. Brandenburg, A.; Lehto, H.J.; Lehto, K.M. Homochirality in an early peptide world. *Astro* **2007**, *7*, 725–732. [CrossRef] [PubMed]
12. Stich, M.; Blanco, C.; Hochberg, D. Chiral and chemical oscillations in a simple dimerization model. *Phys. Chem. Chem. Phys.* **2013**, *15*, 255–261. [CrossRef] [PubMed]
13. Blanco, C.; Hochberg, D. Chiral polymerization: symmetry breaking and entropy production in closed systems. *Phys. Chem. Chem. Phys.* **2011**, *13*, 839–849. [CrossRef] [PubMed]
14. Sajewicz, M.; Gontarska, M.; Kronenbach, D.; Leda, M.; Kowalska, T.; Epstein, I.R. Condensation oscillations in the peptidization of phenylglycine. *J. Syst. Chem.* **2010**, *1*, 651–662. [CrossRef]
15. Danger, G.; Plasson, R.; Pascal, R. An experimental investigation of the evolution of chirality in a potential dynamic peptide system: N-terminal epimerization and degradation into diketopoperazine. *Phys. Chem. Chem. Phys.* **2010**, *6*, 255–261. [CrossRef]
16. Lundberg, R.D.; Doty, P. Configurational and stereochemical effects in the amine-initiated polymerization of n-carboxy-anhydrides. *J. Am. Chem. Soc.* **1956**, *78*, 4810–4812.
17. Bartlett, P.D.; Jones, R.H. A kinetic study of leuchs anhydrides in aqueous solution. *J. Am. Chem. Soc.* **1957**, *79*, 2153–2159. [CrossRef]
18. Mauksch, M.; Tsogoeva, S.B. Life's Single Chirality: Origin of Symmetry Breaking in Biomolecules. In *Biomimetic Organic Synthesis*; Poupon, E., Nay, B., Eds.; WILEY-VCH: Weinheim, Germany, 2011; Volume 2, pp. 823–846.
19. Joyce, G.F.; Visser, G.M.; van Boeckel, C.A.A.; van Boom, J.H.; Orgel, L.E.; van Westrenen, J. Chiral selection in poly(C)-directed synthesis of oligo(G). *Nature* **1984**, *310*, 602–604. [CrossRef]
20. Micskei, K.; Rabai, G.; Gal, E.; Caglioti, L.; Palyi, G. Oscillatory symmetry breaking in the Soai reaction. *J. Phys. Chem. B* **2008**, *14*, 9196–9200. [CrossRef]
21. Sato, I.; Kadowaki, K.; Ohgo, Y.; Soai, K. Highly enantioselective autocatalysis induced by chiral ionic crystals of sodium chlorate and sodium bromate. *J. Mol. Catal. A Chem.* **2004**, *216*, 209–214. [CrossRef]
22. Bonner, W.A.; Bean, B.D. Asymmetric photolysis with elliptically polarized light. *Orig. Life Evol. Biosph.* **2000**, *30*, 513–517. [CrossRef] [PubMed]

Essay

Chirality: The Backbone of Chemistry as a Natural Science

Josep M. Ribó [1,2]

[1] Department of Inorganic and Organic Chemistry, University of Barcelona (UB), 08028 Barcelona, Spain; jmribo@ub.edu; Tel.: +34-934021251 or +34-608551138

[2] Institute of Cosmos Science (IEEC-UB), University of Barcelona (UB), 08028 Barcelona, Spain

Received: 29 October 2020; Accepted: 27 November 2020; Published: 30 November 2020

Abstract: Chemistry as a natural science occupies the length and temporal scales ranging between the formation of atoms and molecules as quasi-classical objects, and the formation of proto-life systems showing catalytic synthesis, replication, and the capacity for Darwinian evolution. The role of chiral dissymmetry in the chemical evolution toward life is manifested in how the increase of chemical complexity, from atoms and molecules to complex open systems, accompanies the emergence of biological homochirality toward life. Chemistry should express chirality not only as molecular structural dissymmetry that at the present is described in chemical curricula by quite effective pedagogical arguments, but also as a cosmological phenomenon. This relates to a necessarily better understanding of the boundaries of chemistry with physics and biology.

Keywords: absolute asymmetric synthesis; biological homochirality; chemical abiotic evolution; chirality; origin of life; spontaneous mirror symmetry breaking; dissipative reaction systems

---

1. Introduction

Cosmological evolution toward the actual world of classical objects passes through many hierarchical stages of increasing complexity (complexity is used here in its definition of interactions forming "complex adaptive systems," which show attributes not expected when their parts are isolated. Such complex systems show adaptive and memory phenomena, dynamic dependence on their initial states, etc.; see, e.g., Gell-Mann [1] and Bak [2]), which implies two symmetry violations (time and charge parity (CP)) and several symmetry breakdowns. As pointed out by Barron [3], "symmetry breaking" is a process leading to a symmetry inferior to that of the initial Hamiltonian but "symmetry violation" defines a process that does not fulfill the physics symmetry conservation theorems [4]. The increase in complexity in the interactions between elementary particles leads to atoms and molecules. The latter are the building blocks of chemical substances and materials, which exhibit a behavior in agreement with our perception of a world constituted by classical objects. The non-classical objects of the elementary particle world reveal phenomena, such as the quantum superposition of states, the double-slit experiment [5], or the experimental demonstrations [6], disabling the so-called Einstein-Podolsky-Rosen paradox of quantum physics [7], which cannot be understood through our sensorial perception. However, the human brain is able to explain, through abstract mathematical reasoning, the world of non-classical objects. The behaviour of objects in the quasi-classical domain (e.g., atoms or molecules) corresponds to the emergence of new properties due to the increase of complexity in the elementary particle interactions. In fact, from a pure quantum field physics point of view, the paradoxical behaviour is not that of the elementary particles but rather that of our classical world. In this regard, the natural sciences are arranged in a hierarchical order (mathematics, physics, chemistry, and biology) where each one is an unconstrained reductionism of the higher one. However, each scientific discipline works with its own terms of complexity [1]. For example, mathematics covers all the natural sciences, but when applied to physics and chemistry it must take into account constraints such as symmetry conservation laws, the symmetry violations occurring in cosmological evolution [3,4], and fundamental thermodynamic principles [8,9], otherwise, nonsensical interpretations can arise.

Chemists work using a methodology based on the following two points:

(i) Consideration of atoms and molecules as quasi-classical objects. In fact, chemistry teaching mostly extrapolated them to atoms as robust classical objects. Atoms and molecules correspond to the frontier, or boundary, of chemistry at its lower level of complexity, that is, to the emergence of the classical from the quantum physics world. This lower frontier of chemistry is well understood by physicists and chemical physicists [10,11] but is generally ignored by chemists.

(ii) Application of the constraints originated by the thermodynamic principles and the description of the behaviour of chemical substances and materials as very large sets of atoms and molecules, i.e., of real chemical samples. Complex chemical systems working in open systems to matter and energy exchange, whose stability derives from the dissipation of energy, are those that represent the upper boundary between biology and chemistry and are still poorly understood.

Despite the above points, chemistry, as a natural science, should study how the complexity of coupled transformations and systems can lead to the replicative and evolutive phenomena of life. The reports on these topics here were mostly restricted to theoretical works [12–16]. However, in the last years several research groups are paying attention to the cooperative role between many reactants in complex reaction networks that has been proposed and even experimentally explored [17–22]. Furthermore, experimental studies in bistable and oscillatory reaction networks in open systems [23–25] have been reported.

The intense and extensive studies in the 20th century on biological chemistry are concerned with the study of the chemical processes of current and actual living beings, but a lack of knowledge of how specific complex adaptive systems determine the emergence of the properties defining life persists. How life's properties emerge from the behaviour of complex adaptive systems may explain why some experts in biological chemistry can believe in creationism theories. An example of the scant interest this has for chemists is shown in the topic of artificial cells, a 21st century topic that is driven by the cooperation between researchers in systems biology and in mathematical physics, but in our opinion it is very unlikely that significant results may be obtained without the collaboration of chemistry. The fault of all this must be placed on the side of the chemical community, which judges the study of chemical evolution as a part of cosmological evolution as a topic of little scientific interest, or as a high-risk research area with low professional rewards. Chemistry today is mostly regarded as a mere applied science, but not as a natural science for the understanding of the world/universe. This problem was already quoted by Hans Primas (ref. [26], pp. 3–4), that chemistry through the overproduction of "specialized" truth is losing the relationship with its neighboring sciences.

Chirality is a well-studied topic in chemistry, but only from a factual point of view. Chemical chirality is mostly treated by pedagogical effective statements, which arise from a factual consideration of the enantiomerism phenomenon, but the *cause* of it is mostly overlooked (Box 1). This has little bearing in the case of applied chemistry, but at the two frontiers of chemistry, one with the elementary particle world and the other with biological processes, chirality appears as a central element. But we have little understanding of its role:

(a) At the molecular level there is the question of enantiomerism; that chemical curricula study chirality as a part of stereochemistry is only related to the existence of structures that break parity, and parity breaking is the consequence of the symmetry violations—time the primordial one—implied in the cosmological emergence of space–time, already dissymmetric [27], and matter. The metaphysics of nature has long ago observed, and been aware of, the chirality phenomenon within the topic of the nature of space [28], well before chemistry was established as a science (see ref. [29] and cites herein).

(b) At the frontier with biology, the dramatic experimental evidence of biological homochirality, overcoming the racemic mixture (racemate), suggests the advantage of chirality for sustaining the information related to the chemical functionalities of biological macromolecules and supramolecular systems implied in replication and in Darwinian evolution, i.e., in the properties that characterize the phenomenon of life.

Box 1. Chemical Chirality.

Chiral dissymmetry is defined by the breaking of the parity operation—simplified as the non-superposition of mirror images—of the molecular structure or of the unit cell of an anisotropic chemical sample. The geometrical representation of these "shapes" is an indirect representation of a chiral electron distribution in space: in molecules, the molecular orbitals; in solids, the electronic bands; in cholesteric liquid phases, the helicoidal solitons, etc.

Dissymmetry of Molecules: Chemists, for example, synthetic chemists working with solutions, use the term "chirality" to refer to the molecular species as defined by a chiral point group [30]. This leads to an incomplete description for isotropic sets of a very large number of units: liquids and solutions of achiral compounds, racemates, enantiopure compounds, and scalemic mixtures.

Dissymmetry of Materials: In the case of crystals, the space point group of the crystal ordering defines chirality when it belongs to one of the 65 Söhncke space groups. For lower anisotropic materials than crystals (e.g., liquid crystals), their symmetry is described by the seven Curie limiting points groups, three of them being enantiomorphic. They also define the dissymmetry of isotropic solutions and liquids: isotropic achiral and racemate samples belong to the same limiting point group, but different from that of enantiopure or scalemic mixtures. See Section 3.

Spontaneous mirror symmetry breaking (SMSB): this can involve enantioselective autocatalytic reaction networks, of high nonlinearity in the enantiomer kinetics, working in systems open to matter or energy exchange. Symmetry changes in SMSB are those of the material's dissymmetry, and correspond to a spontaneous deracemization. Furthermore, trying to explain SMSB on the basis of molecular dissymmetry of molecular reaction mechanisms leads to misunderstandings on the origin of the SMSB phenomenon and how it can be explained (see Section 4).

Physics, in the study of the origin of the universe and of the elementary particle world, necessarily shares questions with the metaphysics of nature. With chemistry at the frontier with quantum physics and with biology, chemistry is also necessarily confronted with metaphysical questions [26,31,32]. However, chemistry as a natural science is concerned with the study of classical and quasi-classical objects, therefore, chemical methodology is freed of any metaphysical considerations [33]. This explains, in our opinion, the general reluctance of chemists to work on topics related to the origin of life that are suspected of being parascientific topics.

Section 1 discusses the role of chirality at the boundary with physics. Section 2 is concerned with the pure chemical domain and discusses the meaning of chemical chirality with respect to the dissymmetry of molecules and chemical materials. Section 3 briefly discusses the conditions to achieve non-racemic stable stationary states instead of racemic ones. Section 4 discusses the role of the scenarios described in Section 3 with respect to the boundary between systems chemistry and systems biology (chemistry/biology boundary). The arguments given in Section 1 are not based on the author's expertise, but on well-established physical concepts (see Acknowledgments). The arguments given in Sections 2 and 3, despite having a well-known physico-chemical basis, are surprisingly mostly ignored by the chemistry community working in chiral synthesis. The discussion in Section 4 is based on reasonable scenarios and some of the statements there are speculative and strongly reflect a personal opinion.

2. Chemical Chirality: Molecular Structures and Chemical Materials

The representation of molecules as classical objects is an approximation that chemistry applies at the borderline between theoretical physics and chemistry (see, e.g., Section 5.6 in ref. [26] and [34–37]). The formation of atoms and molecules is a critical transition due to the increase of complexity in the interactions between elementary particles (Box 2). Totally isolated chiral molecules, as well as the absence of the universal background radiation, will be represented by a superposition of the chiral states that appear as one of the enantiomers through an interaction with the surroundings [26,37].

Box 2. Molecules as Classical Objects.

Molecular structure, therefore molecular dissymmetry, emerges in chemical evolution in the increase of complexity of the interactions between elementary particles. C. F. von Weizsäcker's analysis of physical space (ref [38], pp. 557–560) states that the differentiation between Space and Object of our classical world does not exist in the pure quantum field description. The formulation of the Einstein–Podolsky–Rosen paradox [7] against the quantum mechanical description of elementary particles—today experimentally proven as erroneous—will arise from this attempt to describe the elementary particle world in terms of our emergent space–time world. Therefore, the emergence of our "real" physical space is the emergence of the molecular structure and molecular enantiomorphism: the interactions with the external world leads to wave function collapse, reducing the superposition of states to single-state classical objects.

The understanding of the structure of atoms and molecules is essential in chemistry. As was foreseen by Werner, one of the fathers of chemical science, chemistry today uses knowledge of molecular structure (electronic energy levels and their spatial distribution) to explain the chemical properties of compounds (Figure 1). The use of the geometrical models of molecules to correlate electronic distributions and the properties of chemical substances and materials was successfully introduced in 1947 by Pauling [39], and today this approach is one of the pillars that sustain the study of chemistry. However, pure quantum field mechanisms cannot describe the spatial structure of atoms and molecules. Quantum mechanics requires the collapse of the wave function to describe molecular structures as robust objects, and this occurs because of the quantum decoherence exerted by the interaction with the surroundings and the observer [26,37]: the collapse of the wave function occurs in the irreversible coupling with time as represented by a kinematical superposition of states on a time scale much faster than that of the dynamics leading to thermal equilibrium. Molecular structure would be one of the consequences of the violation of time (reversal) symmetry yielding the thermodynamic arrow of time. Quantum chemistry assumes, in the Born–Oppenheimer approximation, wave function collapse because implicitly it describes a time superposition of states: this allows to calculate molecular structures and molecular properties (e.g., dipole moments). Notice that without wave function collapse, i.e., without decoherence originated by interactions with the surroundings, molecular structure does not exist. In summary, molecular structure can be considered as an emergent "property" arising from the increase of complexity from elementary particles toward classical objects.

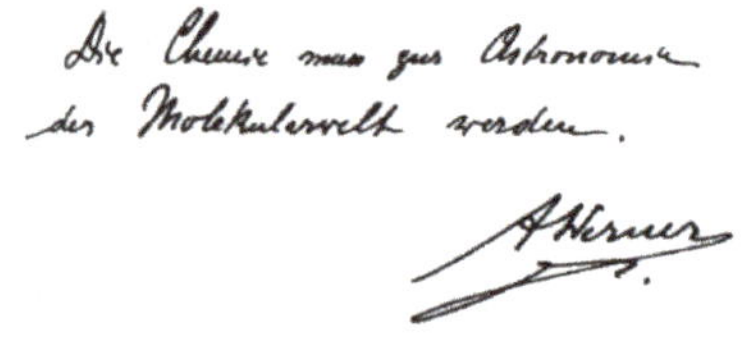

Figure 1. Alfred Werner's autographic quotation "Chemistry must become the astronomy of the molecular world" from "Handschriften Zeitgenössischer Chemiker, gesammelt für die Damenspende des Chemikerkranzes, 1905". We thank K-H. Ernst (ref. [40]) for this historical quotation and the graphic.

Once we assume the existence of molecular structure, the question of enantiomerism needs to solve an unexpected question in order to bring quantum mechanisms in agreement with chemical experience. In the case of chirality, the quantum chemical Hamiltonian does not describe enantiomers as different objects. The wave function of an achiral molecule is represented by only a true ground state, but enantiomers are represented by two strictly degenerate ground states [26,41–44] Under the Born–Oppenheimer approximation, a chiral molecular structure yields a double-well potential (Figure 2) that is invariant with respect to space inversion, and therefore tunneling should be expected, which means that for a large set of molecules spontaneous racemization should be unavoidable. The problem is considered to be solved through the linear combination of ground and excited state (Figure 2, right), but in this case oscillations of the natural optical activity should be expected [45], something that, to best of our knowledge, chemists have never observed. Enantiomer stability requires an increase to infinite height of the barrier (that shown in Figure 2, right) between both minima to avoid tunneling and oscillations of natural optical activity. Many effects have been proposed to bring the quantum chemistry description of enantiomers in agreement with actual chemical experience, i.e., how the barrier between both energy minima might increase to infinity [46]. In respect to the cosmological evolution scenario, the more significant effect is that attributed to the violation of charge parity (VCP) in the weak interaction forming atom nuclei: the Hamiltonian loses the space-inversion symmetry so that the chiral states of the double well are no longer strictly degenerate and will become the eigenstates. Therefore, the minute energy differences between enantiomers arising from the VCP will be the cause of the existence of enantiomers as stable molecular objects and not, as is often proposed in chemistry, as the cause of the bias of the racemic composition in macroscopic samples to yield biological homochirality (see Section 3). Molecular structure and enantiomerism are consequences of the symmetry violations of time reversal and charge parity.

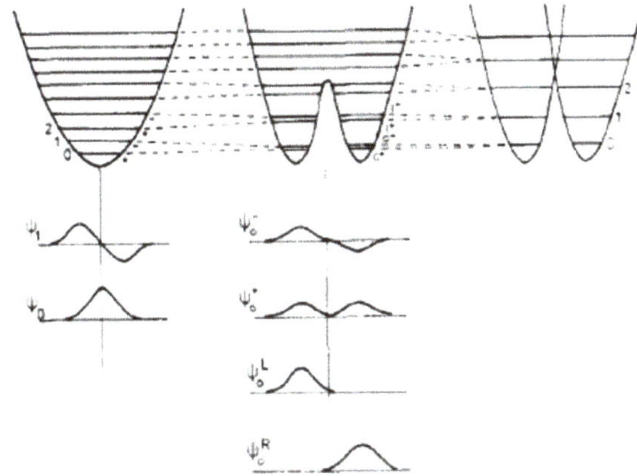

Figure 2. Representation of the vibrational states of a molecule: (**Left**) Achiral molecule; (**Middle**) Chiral molecule, the parity states can be certain when they are mixed, but a certain probability of tunneling leads to chiral oscillations for one molecule and to racemization in a chemical sample. (**Right**) Several effects lead to an infinite barrier separating the two wells (no tunneling), converting the mixed parity states to observable enantiomers such as chemical experience shows. Reproduced with permission from [41]. Reproduced with permission from [41]. Copyright 1979, American Chemical Society.

In summary, despite the fact that quantum mechanics gives a certain probability for tunneling between enantiomers, synthetic chemists, with the experimental evidence at hand, exclude racemization by tunneling because it has never been experimentally detected for carbon atoms as an asymmetric center, and potential chiral amines racemize because of the low inversion barrier to bond angle changes, yielding racemization (in contrast to phosphines): a large set of pure enantiopure molecules should show spontaneous racemization, even in the absence of any racemization reaction.

The Question of Distinguishability between Enantiomers

Enantiomerism is recognized as the breaking of parity. The practical rule to detect it is that the mirror image cannot be superimposed on the original object by translations and rotations. However, this is a factual definition; the causal one should be, as discussed in the following, that homo- and heterochiral interactions are different. In fact, the superposition procedure of the mirror image indicates that homo and heterochiral interactions must be necessarily different, as well as chiral recognition with chiral molecules and physical phenomena arising through the interaction with physical fields showing adequate dissymmetries. Dissymmetry is defined by a set of missing symmetry elements [47] that, for Pasteur, dissymmetry (chemical chirality) corresponds to the absence of any S_n (improper rotation), which obviously implies the absence of symmetry planes and inversion center. Therefore, enantiomerism is classified by the point groups of the molecular structure that break parity [30].

Stereochemistry and organic chemistry handbooks consider enantiomerism/chirality as a singular subgroup of space isomerism. This is a good method for recognizing enantiomeric structures, but underlying the cause of how it is possible that enantiomers, being geometrically identical, must be considered distinguishable particles/molecules. Structures of the same geometry, meaning those defined by historical metaphysical considerations on the nature of space (see [28] and cites therein), lead to the enantiomerism phenomenon only because they show *parity breaking* independently of any other stereochemical argument. The discussion on the existence of "incongruent counterpart objects", i.e., enantiomeric objects, was Kant's (circa 1770) preliminary step in the consideration of space and time as a priori knowledges of human reasoning (see ref. [29] for English translation and discussion of Kant's texts). The a priori perception of space and time would have originated from the chiral nature of our brain's sensorial systems, which, being chiral, allow for the perception of 3D physical space and, being based in dynamical chemical processes, are also bonded to the arrow of time (see Avnir in ref. [48]).

Metaphysical discussions of nature arrived a long time ago at the conclusion that, without chiral recognition, enantiomerism cannot be detected. Discussions on the existence of enantiomeric objects realized the necessity of mutual recognition in order to identify enantiomers: an "incongruent" (chiral) object can only be recognized as such when it can be compared with its counterpart. For example, the Ozma Gedanken experiment describes the impossibility to communicate the chiral sign of an asymmetric object between two cultures located on different planets in the universe [see in ref. [29], pp. 75–95, the Gardner contribution], or also Kant's consideration that two planets having only one handedness, left or right, cannot know if, in the other one-handed planet, the handedness is the same or not (see Kant's considerations on "incongruent objects" (1770–1783) as translated and discussed in [29]). Perhaps human reflections on chirality are as old as the reasoning arising from our sensorial chiral recognition of objects: we may speculate on the first nonwritten reports, but these are graphically expressed in prehistorical hand prints on cave rocks (Figure 3). There the shaman detects a parallel world inside the mountain rock because the being inside the rock shows him the reversed hands to those of the real world.

Figure 3. Prehistorical reflections on chirality? Prehistoric cave right and left printed hands: In the parallel world inside the rock the being shows its reverse hands to the shaman-artist. Copyright 2009, Alamy.

The question of enantiomers as distinguishable particles arose already in the inability of theoretical ideal gas models to include enantiomerism. Lifschitz and Pitayesvski [49], when considering gas kinetics (in a physics textbook!), stated that a pure enantiomer seems not to obey the detailed balance principle (the so-called micro-reversibility principle in chemistry). The detailed balance principle as described by time reversal plus parity should be an even conservation phenomenon, however, for a pure chiral compound, the parity operation leads to a different compound, i.e., to its enantiomer. The detailed balance principle, in the framework of the second principle of thermodynamics, must be necessarily followed; therefore, for the ideal gas model chirality does not exist. Ideal gas conditions assume point molecules and no interactions between them, so that, with such conditions, any differences in the physical properties between ideal gas systems, e.g., heat capacities, can only arise because of differences between molecular masses. Therefore, assuming ideal gas behaviour, the mixtures of enantiomers of different compositions—from homochiral to racemic—should show the same chemical potentials. In the case where we consider non-point-like molecules (i.e., molecules showing extended geometries), but with no differences between the homo- or heterochiral molecular collisions, the heat capacity of homochiral, scalemic, or racemic gas mixtures should be also the same because both enantiomers have the same (average) kinetic energy as well as any other forms of potential energy. Therefore, enantiomeric molecules should be indistinguishable/identical particles. However, in an actual scenario, there are differences between homochiral and heterochiral impacts, interactions, and collisions, and the heat capacities of homochiral, scalemic, and racemic gas mixtures cannot be the same, i.e., the enthalpies of these mixtures are different. Excess enthalpies in enantiomer mixtures may be detected in liquid and dense phases, but in rarified gases and diluted solutions they are under the experimental detection limit. However, despite such an enthalpy difference between homo- and heterochiral collisions being undetectable, there is no obstacle to the contribution of a configurational entropy of mixing ($\Delta S_{mix} = - R \ln 2$ in the racemic composition) to the free energy of the sample, because it is independent of the degree of the interactions leading to chiral enthalpy excesses. This is the Gibbs paradox [50], which describes the configurational entropy of mixing as a non-extensive property (degree of distinguishability) and as a consequence of the distinguishability/identity, or not, between particles. For example, the entropy of mixing, such as pointed out by Schrödinger in 1921 [51], has the same value for a CH_4/CCl_4 mixture as for an isotopical mixture $^{12}CH_4/^{13}CH_4$. The Gibbs paradox has been "solved" by many authors, but probably the best chemical explanation is that of Denbigh and Denbigh [52] that takes into account that the entropy of mixing must be equal and of the opposite sign to that of the reverse process, i.e., the separation of the components of the mixtures through an infinite number of reversible steps (e.g., an ideal isothermic distillation). In consequence, the entropy of mixing corresponds to the value in the infinite limit of reversible separation stages. In this case the uniform convergence of the extensive character of entropy is manifested. This means that distinguishable compounds are those theoretically able to be separated. Therefore, the chemical criterion for indistinguishability is the impossibility of separation. Important is that, in the case of enantiomers, such an enantiomer resolution/separation requires chiral recognition interactions, for example, those of a chiral semi-permeable membrane or of interactions with an enantiopure pure chemical compound or with a chiral force. We can consider that the enthalpic mutual chiral recognition between enantiomers (enantiomeric discrimination $\Delta H_{homochir} \neq 0 \neq \Delta H_{heterochir}$, and $\Delta H_{heterochir} \neq \Delta H_{homochir}$.), i.e., the difference between homochiral and heterochiral interactions, is a causal definition of enantiomerism: the condition for distinguishability, and therefore for the existence of a configurational entropy of mixing ΔS_{mix}, will arise from the enantiomeric discrimination. Notice that all this is equivalent to saying that, without mutual chiral recognition between enantiomers, chemical chirality is not manifested.

In summary, the ultimate chemical definition of chemical chirality should be the existence of enantiomeric discrimination between enantiomers. This has been stated in some chemistry manuals (for example, ref. [53], p. 154), but mostly at the end of the chirality-teaching syllabus. However, the causal significance of enantiomeric discrimination is mostly overlooked; for example, it has been commented that "Homochiral and heterochiral interactions among molecules of like constitution are unlikely ever to be exactly equal in magnitude ($\Delta G_{homo} \neq \Delta G_{hetero}$) because the two types of aggregates are anisometric (diasteromeric). The difference between these interactions was heretofore generally assumed to be nonexistent in the solution or the liquid or gaseous state, or at least too small to be measurable, and of no significant consequence" (ref. [54], p. 3). Notice that without such a small excess enthalpy of mixing, "of no significant consequence", enantiomers would be indistinguishable species and the configurational entropy of mixing, determining the thermodynamic stability of the racemic mixture, would not exist: the entropy of mixing can be considered as the driving force toward the racemic mixture but the cause of this is the existence of an enthalpic discrimination. To forget this has little, if any, consequence in applied chemistry, but may lead to misunderstandings when comparing the stability of non-interacting enantiomers, that is, of pure enantiomers located in different systems and of spontaneous mirror symmetry breaking (SMSB) processes.

In the following we discuss an example of well-established experimental facts pointing at how, in the case of chiral crystals, the existence (or not) of interactions between them reveals (or not) the enantiomerism phenomenon.

The works of Gibbs (1874–1878) on thermodynamic stability, which led to the Phase Rule, were introduced in Europe simultaneously with the experimental verifications of the rule's validity and with the definition of components (C), phases (*P*), and degrees of freedom (F) for different chemical scenarios. The Phase Rule ($F = C - P + 2$) is a general principle governing thermodynamic equilibrium of states completely described by pressure, volume, and temperature. Concerning chirality, a paradox was already detected by van 't Hoff [55,56] when considering a one-component chemical system (achiral or racemizing in the solution) crystallizing into chiral enantiomorphic crystals (the so-called racemic conglomerates, e.g., the case of $NaClO_3$) [57]: when an excess of crystals are in equilibrium with their saturated solution, to be congruent with the experimental behavior of the system, the two enantiomorphic phases must be considered to be thermodynamically identical, i.e., as a unique solid phase (Figure 3). A trivial experimental evidence for this, which to our knowledge is never discussed in physical chemistry text books [58], is that in such a system, pure enantiomorphic crystals or a scalemic mixture of them in contact with its saturated solution, the crystal mixture does not change its enantiomeric excess (ee). For such enantiomorphic solid phases under the experimental conditions of equilibrium, it seems that chemical chirality does not exist. This occurs because in the saturated solution, i.e., at equilibrium, each crystal is isolated from the other ones: here equilibrium is expressed by the corresponding chemical potentials in the interactions between species in the solution and between the species in the solution with the solids, but not between the solids, and when such an interaction does not exist then crystal enantiomorphism cannot be manifested [59]. In summary, in the absence of chiral recognition, enantiomorphic objects are thermodynamically identical objects. This apparent paradox is experimentally observed in the final phase of the procedure of Viedma deracemization [60]. In Viedma deracemization, the wet grinding of racemic conglomerates of one chemical component, yielding enantiomorphic crystals, leads to supersaturation for the bigger crystals because of the higher solubility of the very small ones, and a cyclic process of crystal dissolution and crystal growth is established, which shows a nonlinear enantioselective dynamics leading to deracemization [61]. Any mechanism trying to explain the deracemization process must take into account that there is chiral recognition between the clusters growing to crystals (homochiral interactions occur instead of heterochiral ones). However, when the selective energy input to the system is suppressed, any fluctuation away from thermodynamic equilibrium occurs by monomer (achiral) to crystal (chiral) mechanisms, i.e., there is no chiral recognition, and the enantiomorphic crystals maintain their ee value. Depending on the mechanisms of crystal solution exchange, chirality exists or not and, as manifested by the experimental behavior at the equilibrium, there is only a thermodynamic solid phase (Figure 4).

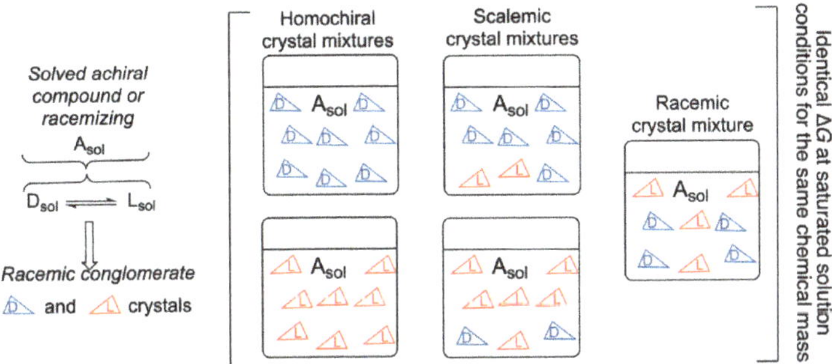

Figure 4. Different crystal mixtures (same crystal sizes) of a system of one chemical achiral component plus solvent that yield enantiomorphic crystals as the more stable mesomorph (crystalline racemic conglomerates, e.g., $NaClO_3$) in equilibrium with their saturated solutions. D/L absolute configuration description is here used instead of IUPAC stereochemical nomenclature. In thermodynamic equilibrium at the same total molar composition, volume, pressure, and temperature, all systems of the figure are thermodynamically identical (crystals are assumed to have the same dimensions, i.e., the same solubility, which experimentally is the case for relatively big crystals of the same size range). The two solid enantiomorphic phases are indistinguishable because at saturated conditions they do not interact with one another, therefore, the enantiomeric excess (ee) of the crystal mixture does not change with time.

3. Molecular Chirality vs. Chirality of Materials

Chemists work with chemical samples and interpret the results with molecular models. For example, the pathways of a synthetic transformation are explained through molecular mechanisms. The physical phenomena arising from the interaction of a material with external forces do not depend on the symmetry/dissymmetry of the molecular structure, but on that of the set of the very large number of molecules composing the material. The properties of a material in its interaction with an external force is given by the Curie principle of symmetry superposition [47,62] determined by the material symmetry elements and the symmetry classes of the external forces [63].

In the case of crystals, for example, racemic conglomerates obtained from compounds that are achiral in solution, the chiral dissymmetry of the solid is given by its classification in one of the 65 Söhncke space point groups. By contrast, chirality in solutions is generally considered by the point group of the molecular components. The latter does not give the right description to interpret the chiral physical response of the solution and may lead to misunderstandings on the meaning of SMSB in enantioselective autocatalysis (see Section 3).

The classical example of the nucleophilic addition to a prochiral carbonyl group (Figure 5) gives us an insight on the differences when the reaction is considered from the point of view of one molecule, or from that of a very large number of molecules. The addition of one molecule can only lead either to the *D*- or to the *L*-alcohol. However, since the reaction paths are reversible for single molecule fluctuations between positive, negative, and zero, then optical activity must occur. The change of the achiral point group C_S, of the prochiral carbonyl to the chiral C_1 of the alcohol, is a decrease of the symmetry order [64] and can be described as a symmetry breaking, but has no sense for a real reacting solution that obviously should lead to the racemate. SMSB refers to the spontaneous formation of a bias from the racemic composition. Notice that the racemate is stable and, in contrast to the single molecule scenario, does not show a chiral response to external forces, and therefore its symmetry (that of the external force) does not break parity.

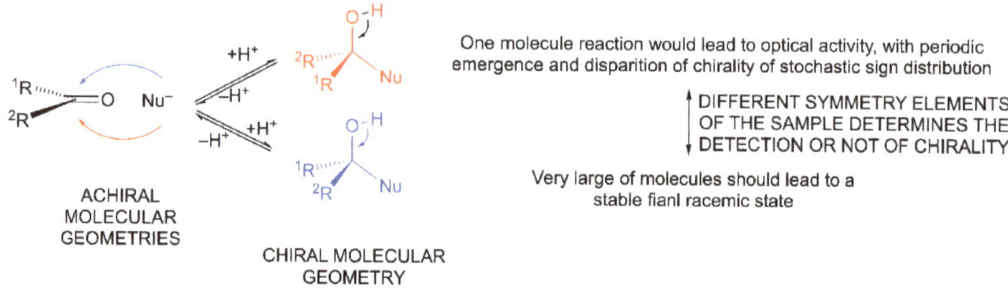

Figure 5. Example of the symmetry decrease of a transformation from achiral to chiral geometry. The consideration of a one-molecule reaction, or of a very large number of molecules, leads to quite different final dissymmetries (see Figure 6).

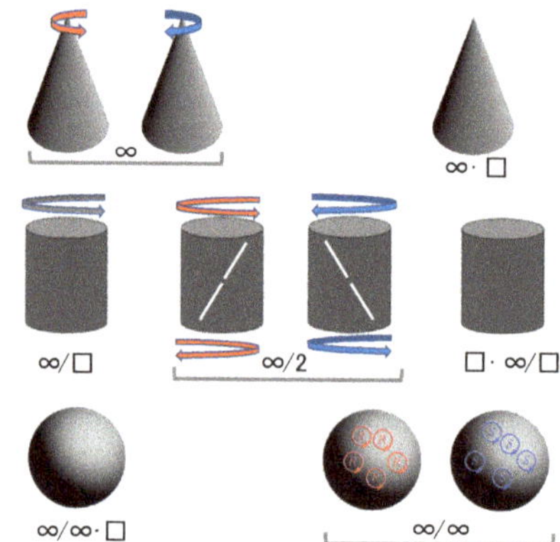

Figure 6. The Curie seven limiting point groups for anisotropic materials other than solid crystals. Enantiomorphism, and therefore chiral physical properties, only can appear for three of these groups. With respect to solutions, their symmetry elements are those described by the spherical groups: the group ∞/∞ describes a solution of enantiopure chiral compounds and also of scalemic mixtures, as, for example, the non-equilibrium stationary states of SMSB (see Section 4). Adapted with permission from [47], Copyright 1988, Elsevier.

Dissymmetry for one molecule is that of the point groups lacking Sn symmetry elements. The symmetry elements of a racemic solution compared to those of the dissymmetry of enantiopure or scalemic solutions are seldom taken into account in organic chemistry, but these can be also described. This is because sample dissymmetry can be described not only by the 65 Söhncke groups, but also by three of the seven Curie limiting point groups [64,65] (see Figure 6), which describe lower anisotropies than those of crystals. The limiting point groups describe not only the symmetry elements of amorphous materials, glasses, liquid crystals, but also ideal liquids and solutions. Limiting point groups able to represent enantiomorphism are the conical ∞, the cylindrical $\infty/2$, and the spherical ∞/∞ groups [47,63]. These are the space groups describing the absence or presence of dissymmetry of solutions and ideal liquids. They also describe anisotropies of oriented molecules in flows, for example, vortices in the conical point groups. The cylindrical limiting groups describe, for example, nematic and smectic liquid crystal phases and also describe the enantiomorphism generated by a shear force at the longitudinal axis of an achiral nematic phase. With respect to molecular solutions, they are described by the spherical groups: $\frac{\infty}{\infty}{\cdot}m$ lacks dissymmetry, i.e., these describe the symmetry elements of achiral and racemate ideal solutions, and ∞/∞ is the space point group of enantiopure solutions and scalemic mixture solutions.

4. Racemates vs. Scalemic Mixtures in Chemical Reactions and Phase Transitions

Obtaining outputs in enantioselective reactions showing biases from the racemic composition is related to the chemical selectivity. Selectivity in the chemical transformation depends on the kinetic and thermodynamic parameters of the reactions, and on the boundary conditions (presence or absence of exchange interactions with the surroundings). A very recent review reported SMSB in crystals, mesophases, and solutions [65]. A brief summary of possible outputs for reversible or irreversible reactions in closed and in open systems is discussed in the following.

Kinetically and Thermodynamically Controlled Reactions

Irreversible reactions or exergonic transformations where the backward reaction rate constant is so small that the return to the initial compounds during the reaction workup may be approximated by zero, as for example,

$$A \rightarrow B;\ (k_B) \qquad A \rightarrow C;\ (k_C), \tag{1}$$

then the selectivity is that given, for initially zero concentrations of B and C, by

$$[B]/[C] = k_B/k_C. \tag{2}$$

In the case of reversible reaction (forward *f* and backward *b* reaction paths):

$$A \rightleftharpoons B;\ \left(k_{fB};\ k_{bB}\right) \qquad A \rightleftharpoons C;\ \left(k_{fC};\ k_{bC}\right). \tag{3}$$

The reaction affinities of (5) at any composition (for ideal solutions) are

$$Af_B = -RTln\left\{\frac{k_{fB}[A]}{k_{bB}[B]}\right\} \qquad Af_C = -RTln\left\{\frac{k_{fC}[A]}{k_{bC}[C]}\right\}. \tag{4}$$

Therefore, at any compositional state the [B]/[C] ratio is given by

$$\frac{[B]}{[C]} = \frac{K_{eqC}}{K_{eqB}} exp\left\{\frac{(Af_B - Af_C\,)}{RT}\right\}, \tag{5}$$

and at thermodynamic equilibrium ($Af_C\ = Af_B = 0$) the selectivity ratio corresponds to the ratio of the equilibrium constants,

$$[B]/[C]\ =\ \frac{K_{eqC}}{K_{eqB}}. \tag{6}$$

In the case of enantiomerism (D and L enantiomers):

$$A \rightleftharpoons D;\ \left(k_f;\ k_b\right) \qquad A \rightleftharpoons L;\ \left(k_f;\ k_b\right), \tag{7}$$

$$\frac{[D]}{[L]} = exp\left\{\frac{Af_D - Af_L}{RT}\right\}, \tag{8}$$

the equilibrium constants and the pairs of reaction rate constants have the same value, therefore at thermodynamic equilibrium

$$[D]/[L] = 1. \tag{9}$$

In systems open to matter and/or energy exchange with the surroundings (boundary conditions), the system behavior cannot be explained solely by the internal reactions, but as a whole [8,66–68], including its interactions with the surroundings. In non-isolated systems, a thermodynamic controlled "equilibrium" final state is in fact a stable non-equilibrium stationary state (NESS). The NESSs show non-zero affinity values and under "thermodynamic control" the species concentration ratios are expressed by Equations (5) and (8). In the case of common selectivity, the ratio [B]/[C] decreases when the boundary conditions drive the system further toward far from equilibrium conditions, but it is always ≠ 1. In contrast, in the case of enantiomerism, all possible NESSs should show racemic composition. The former stable final NESSs are those composing the so-called "thermodynamic branch" of the system. In summary, in the absence of any chiral polarization in enantioselective reactions, at thermodynamic branch scenario, any NESS must be a racemate and, for chemical isomerism and diastereoisomerism, the selective ratio decreases as the NESS moves further away from thermodynamic equilibrium.

However, the central question in energy dissipative systems is that the stability of the NESS does not depend upon whether the energy state functions form a potential well or not, but upon the entropy production (entropy production given by the product of force (affinity) by the current, absolute rate) that it originates [8,9,67,68]. The NESS of the thermodynamic branch becomes unstable at high entropy production values and, under very small fluctuations, the system is driven away from these NESSs toward new states [69].

Common reactions such as (7), which have no nonlinear kinetic dependences, achieve the critical entropy production only at very high affinities and absolute rates, and this occurs for very large species concentrations, where the medium's viscosity is large, leading to diffusion-controlled rates and the breakdown of the mean field assumption, which leads to inhomogeneous distributions of matter and energy [69]. This may lead to ordered structures, called dissipative structures, because their stability occurs thanks to the energy dissipated by the system. However, catalyzed reactions compared to uncatalyzed ones have a higher entropy production because the entropy production rate arises from the product of the force by the current. *Therefore, in complex catalytic networks showing direct or indirect autocatalysis, the critical entropy value of a bifurcation can be achieved also for reactions at ideal solution conditions* [69]: *the NESSs on the thermodynamic branch may become unstable and new stable NESSs can appear*. The system may then transit toward other stable NESSs and may exhibit bistability and oscillatory phenomena, and even chaotic behavior [9].

In the former thermodynamic scenario, autocatalytic enantioselective reaction networks giving rise to SMSB have been theoretically predicted for a long time and were later confirmed by the experimental SMSB examples of Soai's reaction [70,71] and of Viedma's deracemization of racemic conglomerate crystal mixtures [60]. These dramatic experimental evidences of SMSB show how the absence of any chiral polarization leads to a stochastic distribution of the chiral signs [72] between experiments, but under the action of extremely weak chiral polarizations, the perfect symmetric bifurcation transforms into an imperfect (biased) one, leading deterministically to only one of the two chiral signs [73,74]. In the following, we comment on some points in enantioselective autocatalysis in regard to its ability, or not, to exhibit SMSB in open systems.

The irreversible enantioselective autocatalysis of first order,

$$\mathrm{A} + \mathrm{D} \rightarrow 2\,\mathrm{D} \quad (k_{\mathrm{auto}}) \qquad \mathrm{A} + \mathrm{L} \rightarrow 2\,\mathrm{L} \quad (k_{\mathrm{auto}}), \tag{10}$$

in isolated systems may lead to kinetically controlled biases from the racemic composition. For example, in a system that, in addition to (12), contains the direct reaction

$$\mathrm{A} \rightarrow \mathrm{D} \quad (k_{\mathrm{dir}}) \qquad \mathrm{A} \rightarrow \mathrm{L} \quad (k_{\mathrm{dir}}), \tag{11}$$

where $k_{\mathrm{dir}} << k_{\mathrm{auto}}$. When starting at zero concentrations of D and L, the very first ee values, due to the unavoidable stochastic deviations in (11), are transferred to reaction (10) and to the final reaction output [75,76]. Stochastic kinetics simulations show a distribution centered around the racemate of experiments showing ee ≠ 0, and this distribution becomes wider toward higher ee values for higher values of k_{auto}.

However, with respect to SMSB, the interest is in reversible reaction networks. The reversible enantioselective autocatalysis of first order (quadratic in products)

$$A + D \leftrightarrows 2\,D \qquad A + L \leftrightarrows 2\,L, \tag{12}$$

where A is an achiral compound, and D/L an enantiomeric pair of compounds, has a nonlinear dependence on the enantiomer concentrations, but that does not suffice for achieving the critical entropy production for the destabilization of the racemic NESSs of the thermodynamic branch. Only when coupled to other enantioselective reactions can the autocatalytic growth dynamics be sufficiently nonlinear to lead to SMSB [72,77]. Reaction networks that may achieve this are (i) the Frank model [78,79], (ii) Viedma-like deracemizations [60,61], or (iii) enantioselective replicators showing autocatalysis [80,81]. The entropy production analysis and the energy relationships in an open system, where the effect of coupled enantioselective reactions has been studied [68], by simplifying the effect of the coupling of (12) to other enantioselective reactions SMSB, by changing the autocatalytic order ($0.1 \le n \le 2$) in the reaction

$$A + n\,D \leftrightarrows (n + 1)\,D \qquad A + L \leftrightarrows 2\,L, \tag{13}$$

has been recently reported. There [68], it has been shown how enantioselective autocatalysis is able to depart from the thermodynamic branch, showing multi-stability phenomenon and SMSB.

The emergence of bifurcations on the racemic thermodynamic branch, i.e., the stability or instability of the NESSs and the evolution of any non-equilibrium state, can only be understood when the system is considered as a whole, taking into account their coupling between the boundary conditions and the internal reaction network. Reaction coordinate models lose their meaning in open systems: reaction mechanisms describe the racemic thermodynamic branch as well as the scalemic stable NESSs, hence modelling by reaction coordinates and activation energies (Eyring reaction coordinate model) cannot predict the fate in open systems, which are described by the internal entropy production and entropy exchange flows with the surroundings. Moreover, the evolution of any non-equilibrium state is governed by the General Evolution Criterion (GEC) [70,71]: the partial temporal derivative of the entropy production with respect to the forces/affinities must always be negative, and zero at a NESS.

Clearer statements on this would require a detailed discussion and lie beyond the objective of this report. The interested reader may refer to the recent refs. [67,68,82,83] on this and in the specific case of SMSB systems (see Box 3). However, we can summarize as follows. The evolution and stability of the NESSs in open systems is *not* represented by changes in the energy state function of the internal reaction network; stable NESSs are attractors of other compositional states, through thermodynamically irreversible paths determined by a physical potential—non-thermodynamical state function—related to dissipating entropy production flows. Note the contrast with the reaction coordinate model of reversible thermodynamics, which is not a thermodynamic state function.

Box 3. SMSB: Racemic Biases Near Homochirality.

Nowadays, the Soai reaction [84] and the Viedma deracemization [60] are dramatic experimental examples of the previous theoretical reports on the possibility of SMSB in open systems. However, there are extended misunderstandings of this phenomenon.

Irreversible Thermodynamics in Open Systems. In a chemical system, the irreversibility of natural processes (breaking of time reversal invariance) is expressed by the entropy production flows of the internal reaction and by those of the exchange with the surroundings [85]. The increase of the internal entropy production, due to an increase of absolute rates and affinities, may lead to the emergence of new non-equilibrium stationary states (NESSs). In the specific case of SMSB, this leads to the destabilization of the racemic NESS and emergence of stable enantiomorphic NESSs. The physical–chemical description of this is well supported in sound concepts, thanks to the work of physical chemists of the stature, for example, of Glansdorff, Prigogine, and Eigen [8,14,69]. The physical potential governing evolution and stability in these energy dissipative systems is *not* a state function, in contrast to the role of the energy state functions in reversible thermodynamics. The latter allows for the fertile reaction coordinate model that today underlies our understanding and description of chemical reactions. Notice that the reaction mechanism based in mechanics is time reversible. Therefore, it cannot describe or predict the time irreversible SMSB. Notice that the same reaction mechanisms can lead to a racemate NESS or to a scalemic NESS. The entropy production flows (in units of power) originating from the coupling between the internal currents of matter and energy with those of the surrounding interactions [67,81] are the origin of the NESS.

Meaning of Far from Equilibrium Conditions in Reversible and Irreversible Thermodynamics: Reversible thermodynamics assumes the time reversibility along the chemical path, and even in quantum chemical calculations. Here in the isolated system, the evolution of a composition very far from thermodynamic equilibrium can be simulated assuming local equilibrium conditions and a reversible transformation. The reaction coordinate model is based on this, where a hypersurface, defined by the geometrical parameters of atoms or molecules participating in the reaction, is represented as a function of an energy state function, e.g., ΔG. This is a useful, fast, exclusive tool, used qualitatively in current day chemistry and quantitatively in the quantum chemical description of reaction paths. The coordinate reaction model can be approximated for its use in open systems, when the final NESS belongs to the so-called thermodynamic branch. However, this is not the case when new stable NESSs arise as a consequence of the increase of the internal entropy production, such as in the case of SMSB.

Entropy Production vs. State Function Entropy: It is often overlooked that the changes of configurational entropy, for example, when comparing racemates with scalemic mixtures, belong to the energy state function, but that the entropy production term is something rather different, not only in its physical units; it belongs to the heart of thermodynamic principles, i.e., it is the dissipated energy unable to be converted into work, which distinguishes future from past and is something that the reaction coordinate model and the reaction mechanisms do not take into account.

5. Emergence of Biological Homochirality: The Boundary between Systems Chemistry and Systems Biology

Autocatalysis is considered the alpha and omega of the chemical processes supporting life [86], and it is surely significant that enantioselective autocatalysis is a necessary, although not sufficient, condition for SMSB. This link, for example, between biological replicators and chemical spontaneous mirror symmetry breaking [80] is not likely to be fortuitous.

The accepted conceptual proposals on the origin of life are based on cooperative and collective phenomena in the self-organization of autocatalytic networks [12,13,86,87]. This is a scenario of a large set of cooperating small compartmentalized open systems. The complexity in abiotic evolution leads to the emergence of chemical functionalities of supramolecular structures. In this process of complexity increase, there is a non-zero probability for the emergence of catalytic functionalities [86], the specific case of autocatalysis being one of crucial importance for further evolution [14,88,89]. These reaction networks, working in open systems, will be capable of further evolution.

The diversity of chemical compounds that originated in the first stages of chemical evolution is large [90], but the number of organic functional groups, families of organic compounds, reaction types, and reaction mechanisms is not. At the critical transition in the emergence of catalytic reaction networks, only a fraction of the many available chemical compounds belonging to specific homologous families of functional groups would be incorporated into the reaction networks. The role of chirality in the emergence of more specific and effective catalytic functionalities is mostly overlooked, in spite of the fact that many of the biological reactions are not only autocatalytic, but also enantioselective. Enzyme catalysis and the autocatalytic mechanisms of replication of genetic information, both paradigms of life's properties, are supported in dissymmetric supramolecular structures and are enantioselective. These are formed by homochiral building blocks of isomeric diversity, which show the same chiral sign in all living organisms. This points toward a common origin of life's evolution [91] and to its emergence as a collective phenomenon [92].

Living state processes, as well as abiotic chemical evolution on Earth, must be explained by reaction networks with nonlinear kinetics placed in systems unable to achieve equilibrium with their surroundings, such as those of life but within the framework of nonlinear thermodynamics of irreversible processes [9].

The emergence of biases from the racemic composition in chemical evolution has often been explained by appealing to the action of natural chiral forces in kinetically controlled transformations [93–95]. Such a speculation is quite reasonable in astrophysical scenarios, where the ee of chiral primordial organic compounds can be obtained by kinetic control and, because of the very low temperatures, the racemization process would be slow. However, this always implies a decrease with time of the initial ee value. In consequence, this cannot explain the resilience to racemization of the homochirality in the processes of life, which implies low exergonic transformations, i.e., transformations where the approximation of irreversible (one way) reactions cannot be applied and where racemization can occur. However, in more advanced stages of chemical evolution, when enantioselective autocatalytic networks appear, SMSB scenarios are possible.

On the origin of biological chirality: The characteristics of bifurcation dynamics are (a) its stochastic character and (b) its high sensitivity to forces able to act at the bifurcation point. Notice that the stochastic character of (a) changes to a deterministic one, thanks to (b). In this respect, as SMSB processes are extremely sensitive to the chiral polarizations of the surroundings, the biological *common* chiral sign, arising in multiple compartmentalized systems, would occur by the chiral sign selection exerted by a general chiral polarization (a natural physical force or a small ee value of some of the compounds exchanged between systems) of the surroundings. Here, the use of exchange of the compartmentalized systems of reagents, showing small ee values, coming from asymmetric induction reactions in astrophysical scenarios, would determine a common chiral sign for all systems [96]. Notice that such a chiral sign selection by the surroundings, in a stochastic SMSB process, represents a primordial Darwinian selection of the phenotype.

Is Dissymmetry an Advantage for Chemical Evolution?

An important additional question to that of the emergence of biological chirality is if homochirality has some evolutive advantage. In our opinion, it represents an evolutive advantage in two significant aspects, as discussed in the following.

Advantage of dissymmetry in catalysis and replication: Shannon's theory of the measure of information [97,98] widened the concept that information plays a universal role in the relationships between objects. This is a basic concept that can be applied to the natural sciences [64]. Information and evolution have also been discussed from a philosophical point of view (Chapter 5 of ref [38]) and the concepts of singularity (Erstmaligkeit), confirmation (Bestätigung), and noise (Rauschen) that support the potential information of a subject, recognition, and information flows, can be translated to the chemical physics of compounds and chemical interactions and transformations. For example, a special significance is attributed to the information of the form (Gestalt), which in modern chemistry is a self-evident statement, because the geometry of the molecular structure (shape, conformational dynamics) is the description of the electronic distribution and energy that determines the chemical and physical behaviour of the compound and also the informational exchange implicit in its interactions with other molecules and physical fields. In this respect, dissymmetry implies specific informational contents and plays a role in the interaction with other species, depending on whether they are chiral or not.

The inverse relationship of Shannon's potential information with entropy (minimum amount of information when the thermodynamic entropy is a maximum) points to a significant role of configurational entropy when comparing achiral compounds, racemates, and scalemic or enantiopure compounds. The statistical configurational entropy due to indistinguishable spatial arrangements of the molecular structure is expressed by the entropy number (σ) [64]: σ is lower when the symmetry of the molecular geometry decreases. A high symmetry implies a large number of indistinguishable space arrangements, therefore, a high thermodynamic entropy and low potential information. This suggests that an asymmetric structure (C_1; $\sigma = 1$) is the Gestalt for a maximum of potential information. It is surely significant that biological chiral structures are not only chiral, but asymmetric (C_1) (chirality also appears for other point group symmetries [30]). Notice that as biopolymers are composed by homochiral but isomeric building blocks, for example, the α-helix shape of natural L-peptides, they show a higher potential information content than that of a helix composed of a unique type of amino acid (C_2; $\sigma = 2$). Oligomeric enzymes (n-mers) show a higher σ [99], but this is not relevant with regards to achieving a maximum of potential information because the actual catalytic species formed by the interaction with one molecule of substrate (Michaelis–Menten complex) decreases the initial symmetry to C_1. Therefore, the formation of n-mers would be a strategy to increase the number of Michaelis–Menten complexes without modification of the stereospecificity/information exchange in the chemical transformation.

Information theory has been applied to estimate the bits of information contained in the nucleic acids supporting the genetic information (see, for example, ref. [100–102]). This concerns the potential information of the initial compounds, but the reaction path implies the dynamic recognition process between reactants and catalyst and corresponds to the actual information [85] arising from the dynamic interaction between objects (chemical counterparts), and not to Shannon's potential information. Shannon's theory, despite its didactic value, is unable to give quantitative descriptions of the information flows and to express actual information in terms of potential information [85]. Notice that the reaction coordinate can be considered as an expression of the information flow between the interacting subjects along the reaction path [103–105], suggesting prospective work to relate the inverse relationship between Shannon information and thermodynamic entropy, and also between information flows and entropy currents.

Advantage of dissymmetry in electron transport: The chiral biopolymers of metallo-enzymes and nucleic acids are able to transport electrons at large distances through their structure [106,107], from boundary sites to a metallo-prosthetic center in the case of redox enzymes and in the case of nucleic acids, to repair free radicals caused by oxidation damages. The charge transport mechanism in proteins and nucleic acids shows analogies with those of disordered metals: conductivity occurs through some delocalized paths, showing between them energy barriers that are overcome by tunneling and hopping. The emergence of this type of protein has been proposed to be through a self-organizing critical phenomenon [108,109]. The role of these proteins in the emergence of life is crucial because they allow at the interface between systems charge separation, avoiding the recombination of oxidized and reduced counterparts driving metabolic processes. In consequence, such proteins should have emerged at the evolution stage of formation of compartmentalized systems able to exchange chemical energy with their surroundings. In this respect, the group of heme proteins is significant; for example, cytochrome c shows phylogenic traces in its amino acid sequences along many organisms of the phylogenic tree and is commonly used in cladistic studies [110]. In the evolution toward protocells, the ability of electron transport with the surroundings is essential to sustain the primordial internal redox processes.

The role of chirality here was suggested by recent experimental reports on the enantiomeric discrimination between enantiomers and polarized electrons [111–113], which describe the chiral recognition between the helicoidal chirality of a moving electron and, perhaps, on the effect of molecular dissymmetry upon the spin superposition of states of a single electron. The magnitude of the "filtering" effect cannot be explained by the small energy value arising from the spin-orbital effect of the electron at the conducting band of a chiral species [114]. Further work is necessary to understand this dramatic chiral recognition.

An advantage for the function of electron transport in the systems chemistry of life is suggested by the consideration of the two following points:

(1) Electron transport is necessarily a one-electron process, but the redox processes of organic compounds (closed shell) are two-electron redox reactions. The enzymatic machinery of life solves this problem using transition metal prosthetic groups of open shell configuration that receive the single electrons, and, mediated by substrates, is able to show one-electron redox transformations intermediate to achieve the two-electron processes of the redox processes of closed shell organic molecules. In the case of DNA damage, the noxious free radical at the nucleic acid simply reacts with the carried electron to a closed shell configuration.

(2) In regard to the former point, since living systems do not involve solid state reactions, reactions are forbidden when the total electronic spin numbers of the reactants and products are not the same. Therefore, in the case of the primordial one-electron reactions, only free electrons showing the spin sign that agrees with that of prosthetic metal group or of the free radical can participate in the reaction.

Obviously, the last point (2) states that only the half of the electrons, those of the adequate spin value, participate in the reaction. The other half will be scattered and/or lead to undesirable reactions. However, when the dissymmetry information can occur before the electron insulation, or even when the electron spin superposition of states can be selected by the chirality of the biological polymer, an efficient and faster redox process will occur.

6. Concluding Remarks

At the frontier between chemistry and quantum physics, the emergence of molecular structure and enantiomers as stable compounds occurs, thanks to fundamental symmetry violations. At the boundary with biology, chemistry should be able to describe dissymmetric complex systems. From an applied point of view, chemists are beginning to fill the knowledge gap between solution chemistry and complex chemical systems. Discussions and speculations discussed in this report are reflected in Figure 7.

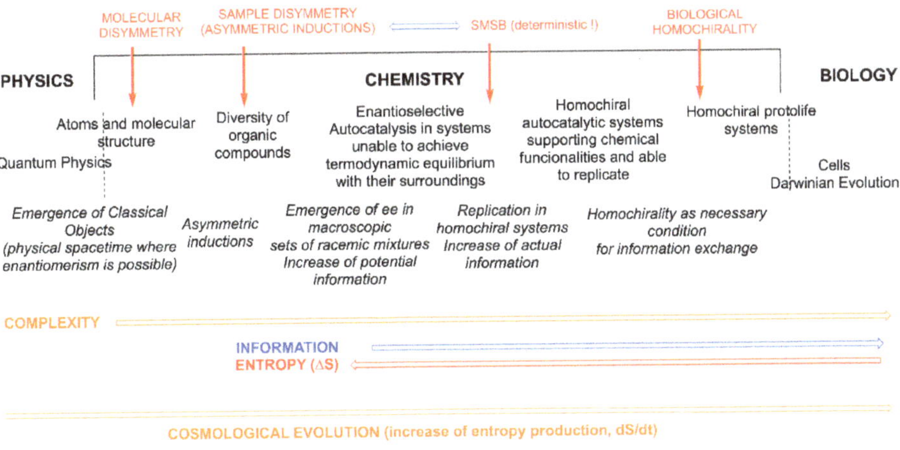

Figure 7. Ordering in the evolution toward complexity, according to the discussion presented in this essay.

Funding: This work forms part of the project CTQ2017-87864-C2-2-P (MINECO).

Acknowledgments: The author thanks P. Seglar for discussion on the parts of the manuscript referring to physics, and to J. Crusats, D. Hochberg, Z. El-Hachemi and A. Moyano for the years-long discussions, whose conclusions are reflected in this essay.

Conflicts of Interest: The author declares no conflict of interest.

References

1. Gell-Mann, M. *The Quark and the Jaguar: Adventures in the Simple and the Complex*; Henry Holt: New York, NY, USA, 1994; ISBN 10 0-8050-7253-5.
2. Bak, P. *How Nature Works. The Sicence of Self-Organized Criticality*; Springer: New York, NY, USA, 1996; ISBN 978-1-4757-5426-1.
3. Barron, L.D. Symmetry and molecular chirality. *Chem. Soc. Rev.* **1986**, *15*, 189–223. [CrossRef]
4. Kosmann-Schwarz, Y. *The Norther Theorems: Invariance and Conservation Laws in the Twentieth Century*; Springer: New York, NY, USA, 2010; ISBN 978-0-387-87867-6.
5. Feynman, P.; Leighton, R.B.; Sands, M. Feynman Lectures on Physics. Available online: https://www.feynmanlectures.caltech.edu (accessed on 29 November 2020).
6. Rowe, M.A.; Kielpinski, D.; Meyer, V.; Sackett, C.A.; Itano, W.M.; Monroe, C.; Wineland, D.J. Experimental violation of a Bell's inequality with efficient detection. *Nature* **2001**, *409*, 791–794. [CrossRef] [PubMed]
7. Einstein, A.; Podolsky, B.; Rosen, N. Can Quantum-Mechanical Description of Physical Reality Be Considered Complete? *Phys. Rev.* **1935**, *47*, 777–780. [CrossRef]
8. Glansdorf, P.; Prigogine, I. *Thermodynamic Theory of Structure, Stbility and Fluctuations*; Wiley-Interscience: London, UK, 1971; ISBN 0-471-30280-5.
9. Kondepudi, D.; Prigogine, I. *Modern Thermodynamics: From Heat Engines to Dissipative Structures*, 2nd ed.; John Wiley & Sons Inc.: Hoboken, NJ, USA, 2014; ISBN 9781118698723.
10. Fein, Y.Y.; Geyer, P.; Zwick, P.; Kiałka, F.; Pedalino, S.; Mayor, M.; Gerlich, S.; Arndt, M. Quantum superposition of molecules beyond 25 kDa. *Nat. Phys.* **2019**, *15*, 1242–1245. [CrossRef]
11. Schätti, J.; Köhler, V.; Mayor, M.; Fein, Y.Y.; Geyer, P.; Mairhofer, L.; Gerlich, S.; Arndt, M. Matter–wave interference and deflection of tripeptides decorated with fluorinated alkyl chains. *J. Mass Spectrom.* **2020**, *55*, e4514. [CrossRef] [PubMed]
12. Kauffmann, S.A. *The Origin of Order. Self-Organization and Selction in Evolution*; Oxford Univesity Press: New York, NY, USA, 1993; ISBN 0-19-505811-9/0-19-507951-5.
13. Kauffmann, S.A.; Johnsen, S. Coevolution to the edge of chaos: Coupled fitness landscapes, poised states, and coevolutionary avalanches. *J. Theor. Biol.* **1991**, *149*, 467–505. [CrossRef]
14. Eigen, M.; Schuster, P. A principle of natural self-organization. *Naturwissenschaften* **1977**, *64*, 541–565. [CrossRef]
15. Eigen, M.; Schuster, P. The Hypercycle—A principle of natural self-organization Part B: The abstract hypercycle. *Naturwissenschaften* **1978**, *65*, 7–41. [CrossRef]
16. Tibor, G. The principles of life (selections). In *The Nature of Life: Classical and Contemporary Perspectives from Philosophy and Science*; Cambridge University Press: Cambridge, UK, 2010; pp. 102–112. ISBN 9780511730191.
17. Krishnamurthy, R. Life's Biological Chemistry: A Destiny or Destination Starting from Prebiotic Chemistry? *Chem. A Eur. J.* **2018**, *24*, 16708–16715. [CrossRef]
18. Krishnamurthy, R. Giving Rise to Life: Transition from Prebiotic Chemistry to Protobiology. *Acc. Chem. Res.* **2017**, *50*, 455–459. [CrossRef]
19. Islam, S.; Bučar, D.-K.; Powner, M.W. Prebiotic selection and assembly of proteinogenic amino acids and natural nucleotides from complex mixtures. *Nat. Chem.* **2017**, *9*, 584–589. [CrossRef]
20. Benner, S.A.; Kim, H.-J.; Carrigan, M.A. Asphalt, Water, and the Prebiotic Synthesis of Ribose, Ribonucleosides, and RNA. *Acc. Chem. Res.* **2012**, *45*, 2025–2034. [CrossRef] [PubMed]
21. Jovanovic, D.; Tremmel, P.; Pallan, P.S.; Egli, M.; Richert, C. The Enzyme-Free Release of Nucleotides from Phosphoramidates Depends Strongly on the Amino Acid. *Angew. Chem. Int. Ed.* **2020**, *59*, 20154–20160. [CrossRef] [PubMed]
22. Patel, B.H.; Percivalle, C.; Ritson, D.J.; Duffy, C.D.; Sutherland, J.D. Common origins of RNA, protein and lipid precursors in a cyanosulfidic protometabolism. *Nat. Chem.* **2015**, *7*, 301–307. [CrossRef]

23. Semenov, S.N.; Kraft, L.J.; Ainla, A.; Zhao, M.; Baghbanzadeh, M.; Campbell, V.E.; Kang, K.; Fox, J.M.; Whitesides, G.M. Autocatalytic, bistable, oscillatory networks of biologically relevant organic reactions. *Nature* **2016**, *537*, 656–660. [CrossRef]
24. Li, Y.; Liu, C.; Bai, X.; Tian, F.; Hu, G.; Sun, J. Enantiomorphic Microvortex-Enabled Supramolecular Sensing of Racemic Amino Acids by Using Achiral Building Blocks. *Angew. Chem. Int. Ed.* **2020**, *59*, 3486–3490. [CrossRef]
25. Sun, Z.; Liu, X.; Khan, T.; Ji, C.; Asghar, M.A.; Zhao, S.; Li, L.; Hong, M.; Luo, J. A Photoferroelectric Perovskite-Type Organometallic Halide with Exceptional Anisotropy of Bulk Photovoltaic Effects. *Angew. Chem. Int. Ed.* **2016**, *55*, 6545. [CrossRef]
26. Primas, H. *Chemistry, Quantum Mechanics and Reductionism*; Springer: Berlin, Germany, 1981; ISBN 0-387-10696-0.
27. Petitjean, M. About Chirality in Minkowski Spacetime. *Symmetry* **2019**, *11*, 1320. [CrossRef]
28. Wehrli, H. *Metaphysics. Chirality as the Basic Principle of Physics*; Copyright Hans Wehrli 2008: Zürich, Switzerland, 2008; ISBN 0521256771978052125677З.
29. van Cleve, J.; Frederick, R. (Eds.) *The Philosophy of Right and Left*; Kluwer Academi Publishers: Dordrecht, The Netherlands, 1991; ISBN 0-7923-0844-1.
30. Mislow, K.; Raban, M. Stereoisomeric relationships of groups in molecules. In *Topics in Stereochemistry, Vol 1*; Allinger, N.L., Eliel, E.L., Eds.; Intersience, John Wiley: New York, NY, USA, 1967.
31. Varela, F.G.; Maturana, H.R.; Uribe, R. Autopoiesis: The organization of living systems, its characterization and a model. *Biosystems* **1974**, *5*, 187–196. [CrossRef]
32. Luisi, P.L. *The Emergence of Life*; Cambridge University Press: Cambridge, UK, 2006; ISBN 10 0-521-82117-7.
33. van Brakel, J. On the Neglect of the Philosophy of Chemistry. *Found. Chem.* **1999**, *1*, 111–174. [CrossRef]
34. Amann, A. The Gestalt problem in quantum theory: Generation of molecular shape by the environment. *Synthese* **1993**, *97*, 125–156. [CrossRef]
35. Bishop, R.C.; Atmanspacher, H. Contextual Emergence in the Description of Properties. *Found. Phys.* **2006**, *36*, 1753–1777. [CrossRef]
36. Joos, E.; Zeh, H.D.; Kiefer, C.; Giulini, D.; Kupsch, J. *Decoherence and the Appearance of a Classical World in Quantum Theory*; Springer: Berlin/Heidelberg, Germany, 2003; ISBN 978-3-642-055576-8.
37. Woolley, R.G. Quantum theory and molecular structure. *Adv. Phys.* **1976**, *25*, 27–52. [CrossRef]
38. von Weizsäcker, C.F. *Aufbau der Physik*; Carl Hanser Verlag: München, Germany, 1985; ISBN 3-446-14142-1.
39. Pauling, L. *General Chemistry*; Dover Publications: Mineola, NY, USA, 1988; ISBN 10 0486656225.
40. Ernst, K.-H.; Wild, F.R.W.P.; Blacque, O.; Berke, H. Alfred Werner's Coordination Chemistry: New Insights from Old Samples. *Angew. Chem. Int. Ed.* **2011**, *50*, 10780–10787. [CrossRef]
41. Barron, L.D. Optical activity as a mixed parity phenomenon. *J. Am. Chem. Soc.* **1979**, *101*, 269–270. [CrossRef]
42. Hund, F. Zur Deutung der Molekelspektren. III. *Z. Phys.* **1927**, *43*, 805–826. [CrossRef]
43. Trost, J.; Hornberger, K. Hund's Paradox and the Collisional Stabilization of Chiral Molecules. *Phys. Rev. Lett.* **2009**, *103*, 23202. [CrossRef]
44. Woolley, G. Natural optical activity and the molecular hypothesis. In *Structures vs. Special Properties. Sturcture and Bonding*; Springer: Berlin/Heidelberg, Germany, 1982; ISBN 978-3-540-11781-0.
45. Harris, R.A.; Stodolsky, L. Quantum beats in optical activity and weak interactions. *Phys. Lett. B* **1978**, *78*, 313–317. [CrossRef]
46. Amann, A. Chirality: A superselection rule generated by the molecular environment? *J. Math. Chem.* **1991**, *6*, 1–15. [CrossRef]
47. Shubnikov, A.V. On the works of Pierre Curie on symmetry. *Comput. Math. Appl.* **1988**, *16*, 357–364. [CrossRef]
48. Avnir, D. On Left and Right: Chirality from Molecules to Galaxies to Rembrandt. Available online: Chem.ch.huji.ac.il/presentations.html (accessed on 29 November 2020).
49. Lifschitz, E.M.; Pitayesvski, L.P. Physical Kinetics. In *Course of Theoretical Physics*; Butterworth-Heinemann Ltd.: Oxford, UK, 1981; Volume 10.
50. Dieks, D. The Gibbs paradox revisited. In Explanation, prediction and confirmation. The phylosophy of science in a European perspective. In *Explanation, Prediciion and Confirmation. The Phylosophy of Science in a European Perspective*; Dieks, D., Gonzalez, W.J., Hartmann, S., Uebel, T., Weber, M., Eds.; Springer: Dordrecht, The Netherlands, 2011; pp. 376–377.
51. Schrödinger, E. Isotopie and Gibbsches Paradoxon. *Z. Phys. A* **1921**, *5*, 163–166.

52. Denbigh, K.G.; Denbigh, J.S. *Entropy in Relation to Incomplete Knowledge*; Cambridge University Press: Cambridge, UK, 1985.
53. Eliel, E.L.; Wilen, S.H. *Stereochemistry of Organic Compounds*; Wiley: New York, NY, USA, 1994; ISBN 978-0-471-01670-0.
54. Kozma, D. *CRC Handbook of Optical Resolutions via Diastereomeric Salt Formation*; CRC Press LLC: Boca Raton, FL, USA, 2002.
55. van't Hoff, J.H. Die Phasenlehre. *Ber. Dtsch. Chem. Ges.* **1902**, *35*, 4252–4264. [CrossRef]
56. Byk, A. Zu den Ausnahmen von der Phasenregel, besonders bei optisch-aktiven Körpern. *Z. Phys. Chem.* **1903**, *45*, 465–495. [CrossRef]
57. Jacques, J.; Collet, A.; Wilen, S. *Enantiomers, Racemates and Resolutions*; Wiley-Interscience: New York, NY, USA, 1981; ISBN 0-471-08058-6.
58. Crusats, J.; Veintemillas-Verdaguer, S.; Ribó, J.M. Homochirality as a consequence of thermodynamic equilibrium? *Chem. A Eur. J.* **2006**, *12*, 7776–7781. [CrossRef]
59. El-Hachemi, Z.; Arteaga, O.; Canillas, A.; Crusats, J.; Sorrenti, A.; Veintemillas-Verdaguer, S.; Ribo, J.M. Achiral-to-chiral transition in benzil solidification: Analogies with racemic conglomerates systems showing deracemization. *Chirality* **2013**, *25*, 393–399. [CrossRef]
60. Viedma, C. Chiral symmetry breaking during crystallization: Complete chiral purity induced by nonlinear autocatalysis and recycling. *Phys. Rev. Lett.* **2005**, *94*, 065504. [CrossRef]
61. Blanco, C.; Crusats, J.; El-Hachemi, Z.; Moyano, A.; Veintemillas-Verdaguer, S.; Hochberg, D.; Ribó, J.M. The Viedma deracemization of racemic conglomerate mixtures as a paradigm of spontaneous mirror symmetry breaking in aggregation and polymerization. *ChemPhysChem* **2013**, *14*, 3982–3993. [CrossRef]
62. Brandmüller, J. An extension of the Neumann–Minnigerode–Curie principle. In *Symmertry. Unifying Human Understanding*; Hargittai, I.B.T., Ed.; Pergamon Press: New York, NY, USA; Oxford, UK; Toronto, ON, Canada, 1986; pp. 97–100. ISBN 978-0-08-033986-3.
63. Newnham, R.E. *Properties of Materials. Anisotropy, Symmetry, Structure*; Oxford University Press: Oxford, UK, 2005; ISBN 0-19-852076-x (pbk).
64. Ercolani, G.; Piguet, C.; Borkovec, M.; Hamacek, J. Symmetry Numbers and Statistical Factors in Self-Assembly and Multivalency. *J. Phys. Chem. B* **2007**, *111*, 12195–12203. [CrossRef]
65. Buhse, T.; Cruz, J.-M.; Noble-Teran, M.E.; Ribó, J.M.; Crusats, J.; Micheau, J.C. Spontaneous deracemizations. *Chem. Rev.* **2021**. accepted.
66. Hochberg, D.; Bourdon García, R.D.; Ágreda Bastidas, J.A.; Ribó, J.M. Stoichiometric network analysis of spontaneous mirror symmetry breaking in chemical reactions. *Phys. Chem. Chem. Phys.* **2017**, *19*, 17618–17636. [CrossRef]
67. Hochberg, D.; Ribó, J.M. Stoichiometric network analysis of entropy production in chemical reactions. *Phys. Chem. Chem. Phys.* **2018**, *20*, 23726–23739. [CrossRef] [PubMed]
68. Ribó, J.M.; Hochberg, D. Spontaneous mirror symmetry breaking: An entropy production survey of the racemate instability and the emergence of stable scalemic stationary states. *Phys. Chem. Chem. Phys.* **2020**, *22*, 14013–14025. [CrossRef]
69. Nicolis, G.; Prigogine, I. *Self-Organization in Non-Equilibrium Systems*; Wiley-Interscience: New York, NY, USA, 1977; ISBN 0-471-02401-5.
70. Soai, K.; Sato, I.; Shibata, T.; Komiya, S.; Hayashi, M.; Matsueda, Y.; Imamura, H.; Hayase, T.; Morioka, H.; Tabira, H.; et al. Asymmetric synthesis of pyrimidyl alkanol without adding chiral substances by the addition of diisopropylzinc to pyrimidine-5-carbaldehyde in conjunction with asymmetric autocatalysis. *Tetrahedron Asymmetry* **2003**, *14*, 185–188. [CrossRef]
71. Islas, J.R.; Lavabre, D.; Grevy, J.-M.; Lamoneda, R.H.; Cabrera, H.R.; Micheau, J.-C.; Buhse, T. Mirror-symmetry breaking in the Soai reaction: A kinetic understanding. *Proc. Natl. Acad. Sci. USA* **2005**, *102*, 13743–13748. [CrossRef]
72. Ribó, J.M.; Blanco, C.; Crusats, J.; El-Hachemi, Z.; Hochberg, D.; Moyano, A. Absolute asymmetric synthesis in enantioselective autocatalytic reaction networks: Theoretical games, speculations on chemical evolution and perhaps a synthetic option. *Chem. A Eur. J.* **2014**, *20*, 17250–17271. [CrossRef] [PubMed]
73. Kondepudi, D.K.; Nelson, G.W. Chiral-symmetry-breaking states and their sensitivity in nonequilibrium chemical systems. *Phys. A Stat. Mech. Appl.* **1984**, *125*, 465–496. [CrossRef]

74. Kondepudi, D.K.; Nelson, G.W. Weak neutral currents and the origin of biomolecular chirality. *Nature* **1985**, *314*, 438–441. [CrossRef]
75. Lente, G. Homogeneous chiral autocatalysis: A simple, purely stochastic kinetic model. *J. Phys. Chem. A* **2004**, *108*, 9475–9478. [CrossRef]
76. Lente, G. Stochastic Kinetic Models of Chiral Autocatalysis: A General Tool for the Quantitative Interpretation of Total Asymmetric Synthesis. *J. Phys. Chem. A* **2005**, *109*, 11058–11063. [CrossRef]
77. Plasson, R.; Kondepudi, D.K.; Bersini, H.; Commeyras, A.; Asakura, K. Emergence of homochirality in far-from-equilibrium systems: Mechanisms and role in prebiotic chemistry. *Chirality* **2007**, *19*, 589–600. [CrossRef]
78. Frank, F.C. On spontaneous asymmetric synthesis. *BBA Biochim. Biophys. Acta* **1953**, *11*, 459–463. [CrossRef]
79. Crusats, J.; Hochberg, D.; Moyano, A.; Ribó, J.M. Frank model and spontaneous emergence of chirality in closed systems. *ChemPhysChem* **2009**, *10*, 2123–2131. [CrossRef] [PubMed]
80. Ribó, J.M.; Crusats, J.; El-Hachemi, Z.; Moyano, A.; Hochberg, D. Spontaneous mirror symmetry breaking in heterocatalytically coupled enantioselective replicators. *Chem. Sci.* **2016**, *8*, 763–769. [CrossRef] [PubMed]
81. Hochberg, D.; Ribó, J.M. Entropic Analysis of Mirror Symmetry Breaking in Chiral Hypercycles. *Life* **2019**, *9*, 28. [CrossRef] [PubMed]
82. Hochberg, D.; Sanchez Torralba, A.; Morán, F. Chaotic oscillations, dissipation and mirror symmetry breaking in a chiral catalytic network. *Phys. Chemisry Chem. Phys.* **2020**. accepted. [CrossRef] [PubMed]
83. Hochberg, D.; Ribó, J.M. Thermodynamic evolution theorem for chemical reactions. *Phys. Rev. Res.* **2021**. accepted.
84. Soai, K.; Kawasaki, T.; Matsumoto, A. Asymmetric autocatalysis of pyrimidyl alkanol and its application to the study on the origin of homochirality. *Acc. Chem. Res.* **2014**, *47*, 3643–3654. [CrossRef]
85. von Weizsäcker, C.F. Evolution und Entropie Wachstum. In *Offene Systeme I*; von Weizsäcker, E.U., Ed.; Ernst Klett-Verlag: Stuttgart, Germany, 1974; ISBN 3-608-93118-X.
86. Kauffmann, S.A. Autocatalytic sets of proteins. *J. Theor. Biol.* **1986**, *119*, 1–24. [CrossRef]
87. Stadler, P.F.; Schuster, P. Dynamics of small autocatalytic reaction networks—I. bifurcations, permanence and exclusion. *Bull. Math. Biol.* **1990**, *52*, 485–508. [CrossRef]
88. Eschenmoser, A.; Volkan Kisakürek, M. Chemistry and the Origin of Life. *Helv. Chim. Acta* **1996**, *79*, 1249–1259. [CrossRef]
89. Hordijk, W.; Hein, J.; Steel, M. Autocatalytic sets and the origin of life. *Entropy* **2010**, *12*, 1733–1742. [CrossRef]
90. Schmitt-Kopplin, P.; Gabelica, Z.; Gougeon, R.D.; Fekete, A.; Kanawati, B.; Harir, M.; Gebefuegi, I.; Eckel, G.; Hertkorn, N. High molecular diversity of extraterrestrial organic matter in Murchison meteorite revealed 40 years after its fall. *Proc. Natl. Acad. Sci. USA* **2010**, *107*, 2763–2768. [CrossRef] [PubMed]
91. Avetisov, V.; Goldanskii, V. Mirror symmetry breaking at the molecular level. *Proc. Natl. Acad. Sci. USA* **1996**, *93*, 11435–11442. [CrossRef]
92. Eigen, M. Selforganization of matter and the evolution of biological macromolecules. *Naturwissenschaften* **1971**, *58*, 465–523. [CrossRef] [PubMed]
93. Avalos, M.; Babiano, R.; Cintas, P.; Jiménez, J.L.; Palacios, J.C.; Barron, L.D. Absolute Asymmetric Synthesis under Physical Fields: Facts and Fictions. *Chem. Rev.* **1998**, *98*, 2391–2404. [CrossRef]
94. Guijarro, A.; Yus, M. *The Origin of Chirality in the Molecules of Life: A Revision from Awareness to the Current Theories and Prespectives of this Unsolved Problem*; Royal Society Publishing: Cambridge, UK, 2009; ISBN 978-0-85404-156-5.
95. Weissbuch, I.; Lahav, M. Crystalline Architectures as Templates of Relevance to the Origins of Homochirality. *Chem. Rev.* **2011**, *111*, 3236–3267. [CrossRef]
96. Ribó, J.M.; Crusats, J.; El-Hachemi, Z.; Hochberg, D.; Moyano, A. Abiotic Emergence of Biological Homochirality. In *Advances in Asymmetric Autocatalysis and Related Topics*; Academic Press: Salt Lake City, UT, USA, 2017; pp. 299–316. ISBN 9780128128251.
97. Shanon, C.E.; Weaber, W.Q. *The Mathematical Theory of Communication*; University of Illinois Press: London, UK, 1949; ISBN 0252725484.
98. Weber, B.H.; Depew, D.J.; Smith, J.D. (Eds.) *Entropy, Information, and Evolution: New Perspectives on Physical and biological Evolution*; MIT Press: Cambridge, MA, USA, 1988; ISBN 0262231328.
99. Comeau, S.R.; Camacho, C.J. Predicting oligomeric assemblies: N-mers a primer. *J. Struct. Biol.* **2005**, *150*, 233–244. [CrossRef]
100. Schneider, T.D. Evolution of biological information. *Nucleic Acids Res.* **2000**, *28*, 2794–2799. [CrossRef]

101. Melkikh, A.V. Quantum information and the problem of mechanisms of biological evolution. *Biosystems* **2014**, *115*, 33–45. [CrossRef]
102. Adami, C. Information-Theoretic Considerations Concerning the Origin of Life. *Orig. Life Evol. Biosph.* **2015**, *45*, 309–317. [CrossRef]
103. Bohme, G. Information und Verständigung. In *Offene Systeme I*; von Weizsäcker, E.U., Ed.; Ernst Klett-Verlag: Stuttgart, Germany, 1974; ISBN 3-608-93118-X.
104. Trambouze, P. Structuring Information and Entropy: Catalyst as Information Carrier. *Entropy* **2006**, *8*, 113–130. [CrossRef]
105. Roderer, J.G. *Information and Its Role in the Nature*; Springer: Berlin/Heidelberg, Germany; New York, NY, USA, 2005; ISBN 10 3-540-23075-0.
106. Regan, J.J.; Ramirez, B.E.; Winkler, J.R.; Gray, H.B.; Malmström, B.G. Pathways for Electron Tunneling in Cytochrome c Oxidase. *J. Bioenerg. Biomembr.* **1998**, *30*, 35–39. [CrossRef]
107. Fink, H.-W.; Schönenberger, C. Electrical conduction through DNA molecules. *Nature* **1999**, *398*, 407–410. [CrossRef] [PubMed]
108. Barraclough, T.G.; Nee, S. Phylogenetics and speciation. *Trends Ecol. Evol.* **2001**, *16*, 391–399. [CrossRef]
109. Vattay, G.; Salahub, D.; Csabai, I.; Nassimi, A.; Kaufmann, S.A. Quantum criticality at the origin of life. *J. Phys. Conf. Ser.* **2015**, *626*, 12023. [CrossRef]
110. Sneath, P.H.A. Cladistic Representation of Reticulate Evolution. *Syst. Zool.* **1975**, *24*, 360–368. [CrossRef]
111. Mondal, P.C.; Fontanesi, C.; Waldeck, D.H.; Naaman, R. Spin-Dependent Transport through Chiral Molecules Studied by Spin-Dependent Electrochemistry. *Acc. Chem. Res.* **2016**, *49*, 2560–2568. [CrossRef]
112. Göhler, B.; Hamelbeck, V.; Markus, T.Z.; Kettner, M.; Hanne, G.F.; Vager, Z.; Naaman, R.; Zacharias, H. Spin Selectivity in Electron Transmission Through Self-Assembled Monolayers of Double-Stranded DNA. *Science* **2011**, *331*, 894–897. [CrossRef]
113. Kettner, M.; Göhler, B.; Zacharias, H.; Mishra, D.; Kiran, V.; Naaman, R.; Waldeck, D.H.; Sęk, S.; Pawłowski, J.; Juhaniewicz, J. Spin Filtering in Electron Transport Through Chiral Oligopeptides. *J. Phys. Chem. C* **2015**, *119*, 14542–14547. [CrossRef]
114. Yeganeh, S.; Ratner, M.A.; Medina, E.; Mujica, V. Chiral electron transport: Scattering through helical potentials. *J. Chem. Phys.* **2009**, *131*, 14707. [CrossRef]

Publisher's Note: MDPI stays neutral with regard to jurisdictional claims in published maps and institutional affiliations.

MDPI
St. Alban-Anlage 66
4052 Basel
Switzerland
Tel. +41 61 683 77 34
Fax +41 61 302 89 18
www.mdpi.com

Symmetry Editorial Office
E-mail: symmetry@mdpi.com
www.mdpi.com/journal/symmetry

www.ingramcontent.com/pod-product-compliance
Lightning Source LLC
LaVergne TN
LVHW070616170726
843515LV00004B/93

* 9 7 8 3 0 3 6 5 0 4 4 2 1 *